Gurdon Trumbull

Names and Portraits of Birds

Gurdon Trumbull

Names and Portraits of Birds

ISBN/EAN: 9783744678353

Printed in Europe, USA, Canada, Australia, Japan

Cover: Foto ©berggeist007 / pixelio.de

More available books at **www.hansebooks.com**

INTRODUCTION.

FOLLOWING is a list of popular and local names applied by English-speaking people to birds which particularly interest gunners: including, however, only those species which are found in the eastern half of the United States; and, again, only those birds which bear aliases to a confusing degree.

One reason that these non-scientific titles have never before been so thoroughly brought together, is a belief that an unravelling of so tangled a skein was practically impossible: so many names being used for more than one species, and so many having been given to one and the same bird. Ornithologists have therefore had the field much to themselves, giving us their long lists of scientific synonyms with little rivalry from the gunners' side of the house.

I believe that the following pages will not only make very conspicuous the difficulties in this branch of our nomenclature, but will show to a great extent what can be done in the premises towards elucidation, and materially simplify the confusion of tongues existing among book-makers, pot-hunters, and sportsmen.

In most cases, where satisfactory identification of species has been arrived at, the names given by early as well as later writers are quoted.

When assigning a name to a locality (without further comment) I have not meant to imply that no other alias of the species is there used, nor that the name is peculiar to the place; but simply that I happen to know of its use in that quarter. Some may feel that I have been over-particular, or unnecessarily explicit in assigning names to individual localities, but I believe that upon a more mature consideration they will thank me for avoiding the temptation to smooth my text by generalizations. I regret that from one cause and another I have not been able to be thus explicit in all cases.

The principal reasons for this multiplication of names are obvious, viz.: differences in size, shape, and color between males and females; periodical changes in plumage; mistaking one variety for another; and, more particularly, differences of opinion as to the names most appropriate.

Many of those English names which perhaps we all ought to adopt, such as "Hooded Merganser," "Hudsonian Godwit," "Bartramian Sandpiper," "Pectoral Sandpiper," etc., are used about as little by the inhabitants of the United States generally as the strictly scientific names; while certain appellations given in our later and best ornithological works, as common among gunners, are quoted from bird-books belonging to a period when popular names were to some extent different from those of to-day. But, though

"Use may revive the obsoletest word,
And banish those that now are most in vogue,"

our gunners have, as a rule, proved themselves a very conservative class, continuing the bird names of their forefathers persistently, despite the teachings and sneers of scientists and book-learned sportsmen. Many of these names, probably, appear now for the first time in print, yet few are of recent origin; and though some may be a little time-worn, they are time-honored, and as familiar in certain localities as "cow," "dog," and "cat." I would remind any who may think it unwise, or idle, to record provincialisms so simple and apparently unmeaning as some of these, that such a view of the subject is itself a provincialism most unreasonable. Names which appear to us absurdly grotesque and outlandish are mediums of communication between men as wise as ourselves, though educated in a different school, and the homely nomenclature of those who shoot, not alone for sport, but for their daily bread, should command respect. It is just now painfully popular to misrepresent and malign the so-ed "pot-hunters;" yet these dear old fellows taught us pretty _uch all we know about hunting, and from them ornithology has gathered its most important contributions.

I have tried to describe the species in as simple English as possible, because I think this freedom from technicalities will be grateful to many. Few, even among our most intelligent college-bred sportsmen, can form a very clear idea of a bird's appearance from the "shop-talk" of scientists, even though provided with a glossary; and it may be broadly stated, with quite a showing of truth, that the descriptions commonly encountered in ornithological works (particularly those of to-day) are only intelligible to those who do not need them.

To further avoid obscurity, I will say that the term "young," as used in my descriptions, refers not to the downy young or the fledglings, but to those birds which have attained, or nearly attained, adult size, and which differ or not, according to their species and sex, from one or both of the old birds.

I will also explain exactly how the measurements "length" and "extent" are obtained. To ascertain the first, lay the bird on its back, hold tip of bill even with edge of table with one hand, pull back the legs with your other hand, and note the point reached by end of longest tail-feather. To ascertain "extent," spread the wings fully out (the bird still on its back), holding tip of one wing flush with edge of table (or other starting-point), and note point reached by the other wing-tip. This measurement has nothing to do with how far the bird itself spread its wings, but shows how far we can spread them, without interfering in the least with construction or natural possibilities.

It may be also well to state here, that the colors of bills and legs given are those of life. In a short time after death (sometimes in a few hours) these colors, particularly those of the bills, change very materially, the brighter hues giving place to a more and more uniform and dusky tint.

The illustrations are by Mr. Edwin Sheppard, of the Academy of Natural Sciences, Philadelphia; and the scientific titles are those adopted by the American Ornithologists' Union, and published in its Check List, 1886.

I have thought that a book which included those birds only in which gunners and sportsmen are interested—pictures of the different species and plumages, descriptions in plain English, full lists of common names, as well as book-names—would, if decently constructed, be a peculiarly intelligible book of reference for those who go gunning. A good picture is worth more for the purpose of identification than all the descriptions ever written, and a picture in simple black and white is in many cases more useful than a colored one, that is to say, for birds whose plumages are entirely different at different seasons, and whose markings and colors necessarily pass through so many intermediate stages. It should be always borne in mind that a bird does not change its plumage as a snake does its skin; that it is impossible to describe each and every variation, and that it is better to leave a great deal to the intelligence of the reader, than to run the risk of hopelessly confusing him by too much detail.

Many will be surprised at the large number of names collected, and some will doubtless wonder why I have omitted this or that name very familiar to themselves. I certainly cannot hope to have gathered all the names of any species, but I feel that I have been a little more than ordinarily careful to keep the lists free from error as far as they go.

I here thank collectively the Eastern gunners who have helped me with material for this work, particularly those living by the Great Lakes, and along our Atlantic coast, with whom I have spent so much time, and formed so many pleasant friendships during the past four years.

NAMES AND PORTRAITS OF BIRDS.

No. 1.

Branta canadensis.

Head, neck, bill, and legs, black; patch about throat, and feathers above and below tail, white. Upper parts of plumage principally brown, this fading into light gray beneath; brown of rump and tail darker, or blackish.

Length a little over three feet ; extent, five feet or more.

Range, as given in A. O. U. Check List: "Temperate North America, breeding in the northern United States and British Provinces ; south in winter to Mexico."

CANADA GOOSE: COMMON WILD GOOSE: BIG GRAY GOOSE: COMMON GRAY GOOSE—Early writers (Hutchins and Hearne) using the latter name for this fowl, but giving that of "Canada Goose" to No. 2, a very similar but smaller bird.

Referred to not infrequently as **HONKER** or **OLD HONKER** in recognition of its hoarse notes, or "honking." At More-head, North Carolina, **REEF GOOSE** (No. 2 being known there as Marsh Goose); and Dresser writes in Birds of Southern Texas, 1865–66: "The shore gunners are well aware of the difference between this [No. 1] and *B. hutchinsii* [No. 2], calling the former the **BAY GOOSE**, and the latter the Prairie Goose."

Early authors tell of its being known at Hudson's Bay as **BUSTARD,**[*] and Sir John Richardson, in Fauna Boreali-Ameri-

---

[*] The bustard of ornithologists belongs to the ostrich family, the Great Bustard (*Otis tarda*) being the largest land bird of Europe.

cana, 1831, speaks of its arrival in the fur countries as "hailed with great joy by the natives of the woody and swampy districts, who depend principally upon it for subsistence during the summer. . . . One goose, which when fat weighs about nine pounds, is the daily ration for one of the Company's servants during the season, and is reckoned equivalent to two snow-geese, or three ducks, or eight pounds of buffalo and moose meat, or two pounds of pemmican, or a pint of maize and four ounces of suet."

No. 1.

In appendix to Townsend's Narrative of Journey Across Rocky Mountains, etc., 1839, it is **BLACK-HEADED GOOSE**; and some writers have termed it **CANADA BRANT**; and in England it has been called the **CRAVAT GOOSE** (Buffon's *L'Oie à cravate*). Yarrell speaks of its being entitled to a place in his British Birds, specimens being so frequently shot "which do not exhibit either in their actions or plumage any signs of having escaped from confinement."

Branta canadensis hutchinsii.

A small variety of our common wild goose No. 1, and in appearance (excepting size) like it in all respects.

Length about twenty-seven inches; extent a little over four feet.

Not common on, or very near to, our Eastern coast, but numerous in the West during migrations. Breeds in Arctic regions.

HUTCHINS'S GOOSE: HUTCHINS'S CANADA GOOSE: HUTCHINS'S BARNACLE GOOSE (the Barnacle Goose proper, *Branta leucopsis*, "casual in Eastern North America," was named from an early belief that it originated in the shell of a barnacle, or, rather, was the natural fruit of a little crustacean): HUTCHINS'S BRANT: LESSER CANADA GOOSE: SMALL GRAY GOOSE: LITTLE WILD GOOSE.

Hearne writes, referring to this variety in his Journey to Northern Ocean, published 1795, "CANADA GOOSE, or PISK A SISH, as it is called by the Indians, as well as the English in Hudson's Bay," and Richardson, in Boat Voyage, 1851, speaks of its being called ESKIMO GOOSE in Rupert's Land.

In Audubon's Ornithological Biography, Vol. III., 1835, we find under the heading of Hutchins's Goose, the following: "In the first article in this volume, that of the Canada Goose, . . . I had occasion to allude to a small species, called by the gunners of Maine the Winter or Flight Goose, which they described to me as resembling the large and common kind in almost every particular except its size. Although it was not my good-fortune while there to meet with the bird spoken of by men who were well acquainted with it, I have no doubt that it is the very species which has been named in honor of Mr. Hutchins." In the "first article," to which Audubon refers, we do not find "Winter Goose," but the other name is introduced as follows: "It is alleged in the state of Maine that a distinct species of Canada Goose resides there, which is said to be much smaller than the one now under your notice, and is described as resembling it in all other particulars. Like the true Canada Goose, it builds a large nest which it lines with its own down; sometimes it is placed on the sea-shore, at other times by the margin of a fresh-water lake or pond. That species is distinguished there by the name of *Flight Goose*, and is said to be entirely migratory, whereas the Canada Goose is resident." Linsley says, in Catalogue of the Birds of Connecticut, 1843: "*Anser hutchinsii*, it is believed, is not unfrequently taken here in the spring, and is called Southern Goose, because it does not winter here." Though this name "Southern Goose" is still remembered in Connecticut, at Stratford, where Linsley wrote, and at Milford as well, the descriptions of the goose to which it belongs, as given by the different gunners, vary very materially; they all agree, however, that the name belongs to a variety smaller than the common wild goose, and very rarely, or never, now encountered.

As these quotations from Audubon and Linsley are both so worded as to leave at least some little room for doubt concerning the local names included, it seems better to give said names just as they appear in the text, without using a more emphatic type.

Giraud writes (1844): "At the eastern extremity of Long Island this species is not uncommon. At Montauk it is known by the name of **MUD GOOSE**."

In an article about common names of wildfowl in Western States (Forest and Stream, May 27, 1886), Mr. J. P. Leach, of Rushville, Illinois, states that the gunners include this with other small geese under the general term "brant," and that this bird is "further distinguished" as **GOOSE BRANT**.

In the neighborhood of Morehead, North Carolina, **MARSH GOOSE**, and on the coast of Texas, **PRAIRIE GOOSE**. (Compare names of this variety with those of No. 1.)

Branta bernicla.

Head and bill, with neck all around, and extreme fore part of body black; on either side of neck a group of white scratches, as indicated in picture. The back, with front of wings, brown,

No. 3.

the feathers paler at their ends; remainder of wings black, or nearly so, as is the tail; the latter, however, being almost concealed by covering of white feathers technically known as tail "coverts." Under parts of plumage grayish brown, the ends of

the feathers touched with white, this producing transverse bars.
Under parts of other specimens, more correctly described as
white, shaded beneath black of fore-breast and along the sides
with ill-defined bars of light brown; in all cases becoming pure
white back of legs.

Length about twenty-four inches; extent forty-six to forty-
eight inches. Legs blackish.

Range, as given in A. O. U. Check List, northern parts of
Northern Hemisphere; in North America chiefly on Atlantic
coast; rare in the interior, or away from salt water.

**BRANT: BRENT: BRANT GOOSE: BRENT GOOSE: BRAND
GOOSE: COMMON BRANT:** has been also called **BLACK BRANT**,
though this latter name is generally applied, and more appropriately, to *Branta nigricans*, a similar but darker bird, rare on our
Eastern coast. The old names "brant," "brent," etc., refer to
the dark color: it is *burnt* or *branded* goose. It ranks high for
table use, and being exceptionally fine when shot late in spring,
the term "May Brant" has long had a momentous meaning
among epicures.

We read in Yarrell's British Birds that "in Shetland it is
called **HORRA GOOSE**, from the numbers that frequent Horra
Sound," and the Rev. Charles Swainson says, in his Provincial
Names of British Birds, 1885: "From the cry of this bird, which
is varied, sounding like the different expressions 'prott,' 'rott,'
and 'crock,' are derived the names **ROTT GOOSE**, or **RAT GOOSE**:
ROAD GOOSE, or **ROOD GOOSE: CLATTER GOOSE** (East Lothian):
QUINK GOOSE: CROCKER." Mr. Swainson also mentions Horra
Goose, and **HORIE GOOSE** as in use at Shetland Isles, and adds
that **BARNACLE** is "the common name for this species in Ireland—a name entirely erroneous. But in some parts the true
Barnacle Goose (*B. leucopsis*) and the Brant are distinguished as
the Norway Barnacle and the **WEXFORD BARNACLE.**"

(See index for other "brant" geese.)

Snow Geese.

Adult. White, with end of wing black; foreparts of plumage frequently stained with reddish brown, this generally more noticeable on front of head. Bill commonly light purplish red, but variable from a more dusky tone to flesh-color, with black

Adult Snow Goose.

"grinning" recess along its sides. Legs deep purplish red, though also variable.

Young. Upper parts bluish gray or lead-color, more or less varied with white; end of wing (or flight-feathers) as in adult. Bill and legs dusky.

Two varieties are recognized by ornithologists, viz.: Lesser Snow Goose, *Chen hyperborea*, and Greater, *Chen hyperborea nivalis*, these being practically alike in form and coloration.

Measurements of smaller bird somewhere about as follows: Length twenty-five inches; extent fifty-two inches. The larger variety: length twenty-seven to thirty-one inches; extent fifty-six to sixty-two inches. The two grading towards one another confusingly.

Another, and less common plumage is that formerly, or at different times, regarded as belonging to a bird distinct from either of the preceding, and again as representing simply a stage in the development of the Snow Goose dress. Birds thus arrayed have been labelled *Chen cærulescens*. This name is

Chen cærulescens.

placed in the "Hypothetical List" of the new Check List,* as "possibly" representing a separate variety. These birds are of the same shape as the snow geese, and are surely most closely allied to them. Head and upper half of neck white; remaining plumage principally grayish brown with more or less bluish gray, the feathers ending paler; edgings of white to back portions of plumage; tone of lower neck dusky; wings plain light bluish

---

* Published by the Am. Ornithologists' Union, 1886.

gray with their flight-feathers ("primaries" and "secondaries")
black or nearly so; rump light gray or more whitish; coloration
of bill and legs about as in previously described snow geese.*

Names of the *whiter* birds, as follows: SNOW GOOSE: WHITE
BRANT (latter name very general in the West): WAVEY or COM-
MON WAVEY of Hudson's Bay region. J. W. Long, in his Amer-
ican Wild Fowl Shooting, speaks of their being known in the
West somewhere as FISH BRANT (an absurdly inappropriate and
libellous designation).

Colonel J. II. Powel writes me from his home in Newport,
R. I.: "I have heard it called MEXICAN GOOSE in this State (I
have killed several here)."† Baird, Brewer, and Ridgway record
RED GOOSE as in use on the Jersey coast (a name mentioned also
in Wilson, 1814), suggested I suppose by color of bill and legs,
and the reddish stains.

These birds visit the Delaware regularly, many of them
congregating near Bay Side, Cumberland Co., N. J., the species
being there known as TEXAS GOOSE.

Names of *Chen cærulescens,* as follows: BLUE GOOSE:
BLUE SNOW GOOSE: BLUE WAVEY: BLUE-WINGED GOOSE:
WHITE-HEADED GOOSE or WHITE-HEAD: BALD-HEADED BRANT
or BALD BRANT.

Though snow geese are rare in most of our Eastern States,
they are exceedingly common in many parts of the West, collecting in countless numbers on the prairies, or transforming
river sandbars into islands of glistening snow. They decoy less
readily than the Canadian and Hutchins's geese, and fly much
higher while passing to and from their feeding-grounds.

---

* Since writing the above, I have become thoroughly convinced that *C.
cærulescens* is a species by itself, distinct from the other geese herein described.

† In Bogardus's Field, Cover, and Trap Shooting, edited by Charles J.
Foster, we read of these birds, with species Nos. 2 and 5, being known as
"Mexican geese" in portions of the West, this term distinguishing them
collectively from the "common wild goose," No. 1.

No. 5.

Anser albifrons gambeli.

Adult. Upper parts principally warm grayish brown, the broad ends of the feathers narrowly edged with brownish white, the pale edgings turned to pure white on tail and certain feathers of the wings; the head and upper neck of closer mixture, or nearly plain brown; extreme front of head (next to bill) white, this white intensified by the brown just back of it, which is of a deeper tint, or blackish. Breast, in high state of plu-

No. 5. Adult.

mage, blackish brown broken only by a few pale or white edgings to the feathers, but, as more often seen, a blotchy mixture of black and white; the feathers of rump, and those beneath tail, pure white. Color of bill varying with different specimens, from flesh color and yellowish, to darker and more reddish tint; the nail at end white or nearly so. Legs and feet orange, the webs lighter, and claws white.

Young. Front of head deep brown instead of white. No black on breast. Bill dull in tone, with nail at end blackish.

No. 5. Young.

Length twenty-seven inches; extent sixty inches.

Its range, as given in A. O. U. Check List, is "North America, breeding far northward; in winter south to Mexico and Cuba."

Though a familiar species to a majority of Western gunners, it is one which we in the East know but very little about.

WHITE-FRONTED GOOSE, or AMERICAN WHITE-FRONTED GOOSE (the latter distinguishing it from European variety *Anser albifrons*): **LAUGHING GOOSE: HARLEQUIN BRANT: PIED**

BRANT. Known in various parts of the West as **PRAIRIE BRANT, SPECKLED BELLY,** and **SPECKLED BRANT,** and very commonly as **BRANT** simply, this being, in other words, *the* brant where hunters are unfamiliar with Eastern bird, *Branta bernicla* (No. 3.), or with Pacific coast variety, *B. nigricans.*

In a letter from Mr. A. B. Pearson, of San Diego, Cal., this species is referred to as the "**YELLOW-LEGGED GOOSE** or **GRAY BRANT,**" and spoken of as "fairly plentiful" there "from November to March."

Anas boschas.

Adult male (in "full" plumage). Head and upper neck brilliant green, with white ring just below; remainder of neck with fore-breast chestnut or chocolate brown. Upper parts of body brown and gray; around tail deep black with greenish

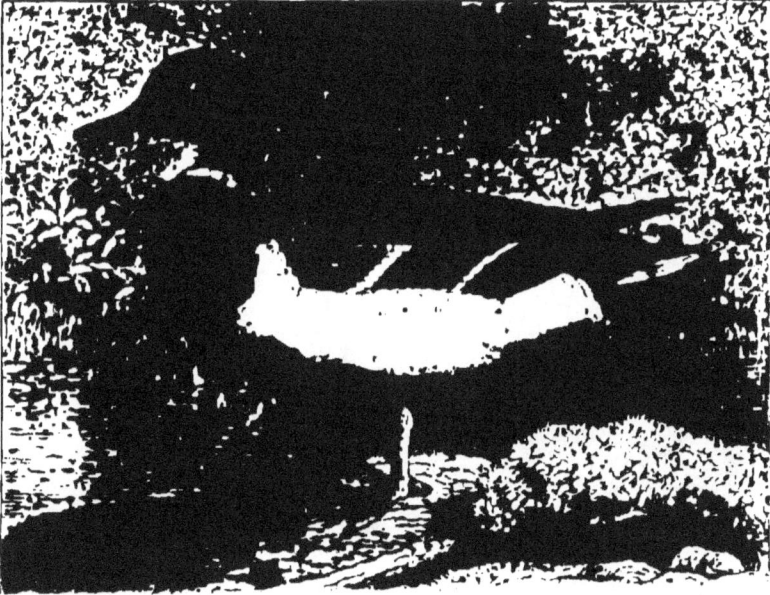

No. 6. Adult Male.

gloss; a tuft of these black feathers turning forward above tail into a little curl. Sides of body white, waved with dusky lines; belly similar, but more grayish white, and very minutely waved. Wings brownish gray with iridescent mark, or "speculum," of purplish blue bordered with white and black. Bill greenish yellow; legs reddish orange.

Size very variable, about two feet in length, and three feet or more in extent. (One old drake now before me spreads nearly forty inches.)

Female. A little smaller than male, with similar wings and feet; bill blotchily marked blackish and orange. General plumage of upper parts dusky brown variegated with light brown or dead-grass color, this latter tint paling to whitish here and there; markings about head and neck fine and streaky; crown

No. 6. Female.

of head dark; throat plain buff; under surface of body varying with different specimens from buff to grayish or soiled white, and spotted with dusky brown.

This is the original of our most common domesticated duck. Though usually shy and suspicious, I have found them feeding with farm-yard cousins in close proximity to barns and dwelling-houses. They are peculiarly ready (male and female) to mate with ducks of other species, and hybrids from these connections are not rare. The offspring of Mallard (No. 6) and Dusky-duck (No. 7) are raised in large numbers at Bellport,

Long Island, by Capt. R. L. Petty, and other old baymen. The
cross was originally obtained from eggs found on a neighboring
marsh, and hatched under a hen. The birds differ greatly in
color, the mixed parentage showing itself in numerous combinations. The female is astonishingly savage during incubation,
flying at one's boot like a mad dog.

Though Mallards visit a few localities in New England
quite regularly (viz., Middleborough Ponds, Mass., vicinity about
mouth of Connecticut River, &c.), New-Englanders, as a rule,
have few opportunities of familiarizing themselves with this
wild stock; and I have heard many gunners tell of losing or
nearly losing good shots at these birds, under the supposition
that they were the property of a neighbor, so closely do they
resemble the barn-yard fowl.

**MALLARD: GREEN-HEAD: WILD DRAKE: WILD DUCK: COM-
MON WILD DUCK.** In English works that treat of wildfowl,
the name **DUCK** alone distinguishes it, other varieties being
referred to as "Widgeon," "Porchard," "Scaup-duck," etc.

At Wilmington, N. C.; Charleston, S. C.; Savannah, Ga.;
and in Florida at St. Augustine, and Sanford, **ENGLISH DUCK:**
and in Louisiana **FRENCH DUCK.**

The female (believed by many a distinct species) is known to
marketmen and others at Detroit, to the "punters" of St. Clair
Flats, at Point Pelee (near head of Lake Erie), West Barnstable, Mass., and in Atlantic Co., New Jersey, as **GRAY DUCK,** *

---

* Though as a rule "Gray Duck" stands in New England for female No. 13
(see both 9 and 13), the name is occasionally borrowed for this less common
species, that is to say, when the latter appears unattended by full-feathered
drake. The two females look near enough alike to be mistaken for each other
by careless observers, and it may be noted that Mr. F. C. Browne, in a list of
"gunners' names" (*Forest and Stream*, Nov. 9, 1876), gives the Mallard's name
"English Duck" as locally applied at Plymouth Bay to No. 13. I was told at
West Barnstable that *their* "Gray Duck" looked "exactly like a common tame
one," and always had "a bright blue spot on its wing." Desiring more evi-
dence, I asked that the first one killed might be sent me, and a few weeks
later received a gray Mallard, it being, I was told, one of three seen.

this name being used in like manner on the Niagara by some of the gunners, though applied more commonly there to No. 13.

In Baltimore, the female is **GRAY MALLARD**, the marketmen furnishing their patrons with three varieties of mallard, as follows.: " green-head," "gray mallard," " black mallard," the last-named variety being *Anas obscura*, No. 7. The name " gray mallard " is also commonly used (for No. 6) at Washington, D. C., and Alexandria, Va., though generally in these localities to include the full-plumaged drake as well.

The species is several times referred to by Lewis and Clarke, 1814, as **DUCKINMALLARD**. If this word occurred but once it might be considered a typographical error, but it certainly seems to have been so printed intentionally. Old writers commonly referred to this fowl as the " duck and mallard." Bartram, for example, in his Travels through North and South Carolina, etc., 1791, speaks of " the great wild duck, called duck and mallard ;" not meaning duck *or* mallard (though, as previously stated, the single word " duck " sometimes distinguishes this from all other species) but duck and *drake*—mallard being derivatively *male*. The above queer name is therefore believed to have grown from this old custom.

At Hudson's Bay, according to Fauna Boreali-Americana, 1831, **STOCK DUCK**; and we find the following in Rev. Charles Swainson's Provincial Names of British Birds, 1885 : "**MIRE DUCK** (Forfar) ;" "**MOSS DUCK** (Renfrew, Aberdeen);" "**MUIR DUCK** (Stirling) ;" and two names already mentioned, as follows : " Gray Duck (Lancashire, Dumfries);" " Stock Duck (Orkney Isles)."

Anas obscura.

Prevailing color blackish brown, of lighter shade below; head and neck also lighter and more grayish in tone. Wing-mark, or "speculum," purplish blue changing to green, and bordered with black; lining of wings white.

Size very variable, but about that of species immediately preceding, and of same general shape.

Bill greenish yellow, with black nail at end. Legs orange red, the webs dark.

No. 7.

Found chiefly in Eastern North America; "West to Utah and Texas;" North to Labrador.

DUSKY DUCK: DUSKY MALLARD: very generally known in New England and Middle States as **BLACK DUCK** (see Nos. 28, 29). As we move westward, and farther south, we hear **BLACK MALLARD**; and reaching South Carolina or Georgia, **BLACK ENGLISH DUCK**, latter title continuing into Florida (though the name Black Mallard is also heard in the far South).

2

No. 8.

Anas americana.

Adult male. Forehead and crown white, or nearly so; remainder of head, with upper neck, pale buff, profusely speckled with black or greenish black; a large patch of glossy green beginning at the eye and sweeping backward; the speckling of head breaking into this green, or blending with it to greater or less extent; the throat nearly plain buff, though dusky immedi-

No 8. Adult Male.

ately back of bill; lower neck (all around), sides of body, and shoulder feathers light brownish red, with more or less pinkish cast; this color almost uniform about front and sides of neck, waved with dusky lines on sides of body, more grayish and waved also with dusky lines at back of neck and on the shoulder feathers. Back gray, waved minutely and obscurely with lighter lines, this coloration turning to a more distinct pattern of dusky and white lines over base of tail; the tail itself chiefly brownish gray. Fore part of wing, with broad field of white; back of this white, a glossy green wing-mark, or speculum, bordered with black; the lower halves of inner secondaries (long feathers starting at first joint from body) black; remainder of wing chiefly brownish gray, with a few feathers edged or tipped with white. Breast and belly continuously pure white; the feathers immediately beneath the tail black, this black continued upward a little over root of tail, the white of lower parts being likewise continued upward at sides of rump. Bill light bluish gray with black tip. Legs and feet bluish gray with darker webs.

We often find these drakes with green head-patch very imperfect, with little or no greenish gloss to speculum; the latter showing, instead, as a deep brown or blackish space; top of head considerably speckled; white of wings pervaded to greater or less extent by gray, and brownish red at sides of body, with very few of the wavy markings above mentioned.

No. 8. Female.

Female. Head and neck streakily speckled with dark brown and grayish white, changing to brown and light yellowish brown on lower neck, fore-breast, and sides of body; the darker brown markings becoming less positive and almost disappearing along the sides. Upper parts chiefly grayish brown; the feathers of back and shoulder region edged with light brownish buff; wing-mark, or speculum, blackish, with slight greenish gloss, and bordered in front with white. Under surface of body nearly white, the feathers immediately beneath the tail deep brown. Bill and legs much as in the male.

Young male (before beginning to assume dress of old drake), very similar to female just described.

Length eighteen to twenty-one inches; extent thirty to thirty-four and a half inches; bill (measured along the side) one and five eighths to one and three quarter inches.

Found throughout the whole country during migrations.

WIDGEON (see Nos. 9, 12, 13, 17, 31): more correctly the **AMERICAN WIDGEON** (distinguishing it from Old World widgeon, *A. penelope*): **BALD-PATE** (see No. 29): **GREEN-HEADED WIDGEON**.

I find the name "Widgeon" in common use at North Scituate, Mass.;* on Long Island at Moriches, Bellport, and Seaford (Hempstead); throughout New Jersey; at Norfolk, Va.; in the neighborhood of Chicago, and at Hennepin on the Illinois River.

We also hear "Bald-pate" at Chicago, and at Havre de Grace, Md.; and **BALD-HEAD** at Hennepin, and at Savannah, Ga.

In Massachusetts (when encountered) it is known at Provincetown as **SOUTHERN WIDGEON**; at North Plymouth as **CALIFORNIA WIDGEON**; and in the vicinity of Edgertown as **WHITE-BELLY**. This latter name is a familiar one also to the older gunners of Milford, Conn.

To some of the gunners of Detroit it is the **POACHER**, being so called from its well-known habit of foraging upon the food for which other ducks have dived.

At Washington, D. C., Alexandria, Va., and Morehead, N. C., **BALD-FACE** (not recognized in latter locality by any other name).

---

* It is not a common duck in New England, and I do not remember hearing it referred to by any local name in Maine or New Hampshire.

Lawson, in his New Voyage to Carolina, 1709, says: "The bald, or white faces are a good fowl; they cannot dive, and are easily shotten."

At Crisfield, Md. (east shore of Chesapeake), and Wilmington, N. C., **BALD-CROWN**: at St. Augustine, Fla., **BALD-FACED WIDGEON**.

Dr. David Crary, of Hartford, Conn., tells me that while shooting in Benton Co., Oregon, in 1885, he found this species in enormous flocks on the wheat-fields, and that it was there called the **WHEAT-DUCK**.

Robert Kennicott (cited by Baird, Brewer, and Ridgway) speaks of its being known to *voyageurs* throughout the Fur Countries as **SMOKING-DUCK,*** and Pennant, in his Arctic Zoölogy, 1785, tells of its being "sent from New York, under the name of the Pheasant Duck;" but the latter name (as others have suggested) was probably applied by mistake.

---

* Probably because its note was thought to resemble the puffing sound made while smoking.

No. 9.

Anas strepera.

Adult male. Head and neck pale buff gray, freckled with grayish brown; crown and hind neck darker; lower neck and breast dark slaty-brown, with scale pattern of nearly white

No. 9. Adult Male.

lines. Union between lighter neck and darker plumage beneath commonly more or less abrupt, and sometimes at this juncture there is a well-defined black ring; this ring, though seldom perfect, is found in various stages of imperfection.* Belly white,

---

* I have found this ring nicely developed upon three or four different

with faint touches of gray; feathers immediately beneath tail black. Fore part of back and sides of body slaty-brown, pencilled with wavy lines of dull white; lower part of back darker, and becoming black on rump and above tail. Certain of the long feathers sweeping back from shoulder region tinged with pale yellowish brown; tail feathers and much of wing brownish gray. Speculum (*i. e.*, outer end of feathers growing from second bone of wing) white, an edging beneath and broad patch in front of speculum black; in front of this black a patch of maroon, or dull mahogany color.

Bill blackish, about one and seven eighth inches long. Legs dull orange yellow.

No. 9. Female.

Female. Head and neck much like male, though a little more distinctly spotted; the finer markings of neck mingling with those of breast without abruptness. Speculum and black

drakes, two of which I sent to the United States National Museum for Mr. Ridgway's inspection. He kindly writes me (Dec. 5, 1885): "It is probable that they represent an 'individual' variation of plumage, probably a very high state of plumage. I find indications of the black collar in specimens belonging to our collection, but in none is it so strongly marked as in your specimens." I have found no reference whatsoever to this marking in ornithological works.

next it also similar to drake; but the black more limited, and the maroon tint but slightly indicated or absent. Under surface of body white, spotted more or less completely with brown. Breast and all remaining parts of plumage (excepting certain plain brown wing feathers) mottled with dusky grayish brown and light yellowish brown. Legs pale yellow. Bill dusky brown above, with edges and under part yellow.

Length about twenty-two inches; extent thirty-four to thirty-five inches.

"Nearly cosmopolitan. In North America breeds chiefly within the United States" (A. O. U. Check List).

GADWALL (spelled also "Gadwell," "Gadwale," etc.): **GRAY DUCK** (see Nos. 6, 13). These two are its book names. The first mentioned,which is pre-eminently "booky," I find used at Chicago even by marketmen and gunners; and the latter name at Chicago, on the Illinois River, and by some at Savannah, Ga.

Though rather a rare visitant on Long Island, it is known (when it does appear) at Moriches as **SPECKLE-BELLY**, and at Seaford (Hempstead) as **CREEK DUCK**; the latter being a common name also at Morehead, N. C., and in the vicinity of Savannah.

On the coast of New Jersey, at Barnegat, Tuckerton, and Atlantic City, it has long been known as the **BLATEN DUCK** (blatant, or bleating, like "*strepera*," from its obstreperousness); and Giraud (1844) speaks of its being called "**WELSH DRAKE** or **GERMAN DUCK**" at Egg Harbor. I have made numerous inquiries for these last two names among the Jersey coast duckers, but have found no one who remembered having heard either of them.

At Wilmington, N. C., and at Savannah, **WIDGEON** (see our widgeon of the books, No. 8; also Nos. 12, 13, 17, 31). Called also at Savannah **GRAY WIDGEON** (see No. 13); four aliases being used, therefore, in this locality: Gray Duck, Creek Duck, Widgeon, and Gray Widgeon.

I am told by S. E. Topping, of Moriches, that this duck is known in Mobile as the "Chickcock;" and Captain Robert L. Petty, of Bellport, tells of hearing it called "Chickacock" at

New Orleans; two forms of one and the same name, doubtless.
These two Long Island baymen are not only thoroughly reliable,
but they are peculiarly well acquainted with our water-fowl; and
my only reason for not using more emphatic type is my ignorance concerning the orthography.

Though occasionally met with in New England, I have heard
no local name applied, and indeed have nowhere found the Gadwall common on the Eastern coast.

Though abundant in certain interior localities, this is not
(taking the country through) a numerous species, as compared
with other varieties.

Since writing the above, Rev. Charles Swainson's Provincial
Names of British Birds has appeared, and in it we find "SAND
WIDGEON (Essex)," and the name RODGE with no locality
assigned.

No. 10.

Anas carolinensis.

Adult male. Above principally grayish; most of head and upper neck rich reddish brown, or "chestnut bay;" a green patch, blackened at lower border, surrounding eye, and sweeping backward and downward to black tufty feathering of nape; at lower edge of patch, and beneath the eye, an indistinct brown-

No. 10. Adult Male.

ish white line; bill black, or nearly so. Neck, beneath its reddish brown portion, with much of upper plumage and sides of body, delicately waved with lines of white and black. Iridescent wing-mark, or " beauty spot," green, framed with black, buff, and white; and above this a few feathers (starting at inner frame-work of wing) marked lengthwise with black and line of

white.   Breast light buff, with blackish spots; at either side of
breast a white bar; belly white, though often with buff tinge;
feathers beneath tail black, with light buff patch at each side.
Legs bluish gray, tinged with flesh color.

Female.   Principally dark brown and light yellowish brown
or buff ; the lighter color noticeable upon edges of the feathers,
but more closely mixed in a streakily speckled manner about
neck and head; crown of head and streak running back through
eye, dark; throat pale buff; on lower fore neck, breast, and

No. 10.   Female.

along upper sides of body the buff tint predominating, and of a
rather deeper shade, and the dark markings more spotty.   Under
surface of body white, with dusky markings back of legs.   Wings
nearly as in male.

Length about fourteen inches; extent twenty-two to twenty-
four inches.   Bill narrow, and nearly same width throughout.

Range, North America at large.

GREEN-WINGED TEAL, or GREEN-WING, simply, so called
very generally : AMERICAN GREEN-WINGED TEAL, distinguishing it particularly from European Green-wing, *Anas crecca,*
which strays to us occasionally ; and termed likewise by certain

early writers, the **AMERICAN TEAL*** and **LEAST GREEN-WINGED TEAL.**†

At Bath, Me. (to the older gunners at least), **MUD TEAL;** at Moriches, Long Island, **WINTER TEAL** (see No. 31); at Morehead, N. C., **RED-HEADED TEAL.** No other name for the species is recognized in latter locality, so far as I can discover (1884), and this is certainly a more distinctive appellation than that of " Green-wing." Audubon says : " Its general name, however, is the ' Green-wing;' and a poor name in my opinion it is, for the bird has not more green on its wings than several other species have." And he adds: " Very many birds are strangely named, not less in *pure* Latin, than in English, French, and Dutch."

---

* Latham.   † Bartram.

No. 11.

Anas discors.

Adult male. Crown of head and chin black; a white black-edged crescent between bill and eye; remainder of head, with a little of the neck adjoining, bluish lead color, with lavender tinge; bill black. Upper plumage principally dark brown with black, and spots, bars, and streaks of buff; front of wing sky

No. 11. Adult male.

blue; blue also on some of the longer shoulder-feathers; wing-mark, or speculum, green, with white band between it and above-mentioned blue, and a narrow line of white on opposite border of speculum.

Lower plumage light brown or reddish buff, thickly speckled with black; this marking changed to waved bars upon the flanks; at either side of tail (upon the rump) a patch of white; feathers immediately beneath tail black. Legs dull yellow.

Female. Quite differently dressed from drake, excepting wing, which is similar. Most of upper parts dark brown, with narrow pale buff edgings to the feathers; crown of head and streak running back of eye dark brown; sides of head and upper

No. 11. Female.

neck finely and streakily speckled with dusky brown upon a white ground; throat white. Lower neck and breast, with sides of body, pale buff, mottled with brown; the buff tint fading to dingy white on belly, where the markings are smaller and less distinct.

Length, fifteen to sixteen inches; extent, twenty-six to thirty inches.

Range, North America at large, principally east of Rocky Mountains.

BLUE-WINGED TEAL, or **BLUE-WING** simply; so termed generally: also known as **SUMMER TEAL.** This latter name is common at Moriches, Long Island, and I am inclined to believe that I have heard it among the gunners of other localities, but this is the only note I have.

Called by early writers **WHITE-FACED TEAL,** and **WHITE-**

FACED DUCK; and Dr. Woodhouse, in Sitgreaves' Expedition— Zuni and Colorado Rivers, 1853, calls it **COMMON BLUE-WINGED TEAL**; thus distinguishing it from "Western blue-winged teal," *Anas cyanoptera;* the latter, however, being better known as "Cinnamon Teal."

No. 12.

Aix sponsa.

Adult male. Head and plume principally dark green and purple, the green predominating above, the throat white; a line of white running back from bill over eye, another behind the eye; both these white streaks continued along the crest; white

No. 12. Adult Male.

of throat branching upward as shown in picture. Lower neck and the breast purplish chestnut, or maroon, with triangular white spots. Belly white. Sides delicately waved with fine lines of straw color and black, and separated from maroon of

breast by conspicuous white and black bar; the feathers covering flanks barred at their ends with white and black, and a patch
of purplish chestnut at either side of rump. Upper parts generally dark, of brownish and brownish gray tone, varied with velvet black, brightened with greenish bronze; the wings broadly
glossed here and there with purplish blue and other iridescence,
and narrowly edged behind (on ends of the broad blunt feathers) with white; hairy filaments, varying in tint, at either side
of tail. Bill, with ridge, tip, and under part, black; its sides
red near head, and whitish farther forward. Eyes red. Legs
yellow.

Female. A rather quietly attired, principally grayish and

No. 12. Female.

slaty-brown duck, though with considerable iridescence. Bill
dusky; a narrow edging of white next it. Crest slight and
scarcely noticeable. Throat, patch around eye, and the belly,
white. Breast tan brown, streakily marked with pale buff;
sides also brown, with spots of dull white. Legs yellowish
brown.

Length eighteen to nineteen inches; extent twenty-eight to
twenty-nine inches.

"Temperate North America, breeding throughout its range"
(A. O. U. Check List).

3

WOOD DUCK (see No. 22): **SUMMER DUCK**: widely known by one or both of these titles, and commonly conceded to be the most beautiful of our water-fowl. It may be added that as a table bird it stands also very high.

At East Haddam, Conn., it is the **WIDGEON.*** "A good many here call it Wood Duck," said a local gunner, "because it builds its nest in trees, but most of us know that its real name is Widgeon." Farther down the Connecticut River, we hear **WOOD WIDGEON**: "Always called it so," said an Essex ducker, "until Clark told us its right name." Mr. John N. Clark, of Saybrook, near by, being the authority referred to.

At Pocomoke City (Worcester Co.), Maryland, and in the vicinity of Charleston, S. C., **ACORN DUCK**. Mentioned in Belknap's History of New Hampshire, 1784, as **CRESTED WOOD DUCK**; and Latham writes, Synopsis, 1785: "By some called **TREE DUCK** (see No. 22). Our "Tree-ducks" *proper*, met with along southwestern border of the United States and southward, belong to the genus *Dendrocygna*.

---

* See our Widgeon of the books, No. 8; also Nos. 9, 13, 17, 31.

No. 13.

Dafila acuta.

Adult male. Head and upper neck rich brown with copper-red reflections; portions of hind neck black; lower hind neck, front of back, and sides of body evenly waved with dusky gray and white; much of wing plain gray and grayish brown; wing-

No 13. Adult Male.

mark, or speculum, green, changing to copper-red, and edged with white, cinnamon, and black; the tapering feathers starting at inner framework of wing, and sweeping along the lower back, are black centrally, with broad gray, white, or brownish margins.

Tail gray, with black about its roots, its elongated central feathers black, or nearly so. Front of neck, and under parts generally, white, tinged more or less with yellowish or rusty stain; the white of neck branching into the dark color above, as shown in picture. Bill black, turning to bluish gray along its sides. Legs and feet bluish gray.

Measurements about as follows: length twenty-nine inches (but governed, of course, greatly by variable development of central tail-feathers); extent thirty-six inches; bill, measured along top (from feathering at base) two to two and three-sixteenth inches, and narrow, with sides nearly parallel.

No. 13. Female.

Adult female. Plumage very different from full-dressed drake: a quietly clothed "gray duck." Central feathers of tail projecting but slightly beyond those next them. Upper plumage principally dark grayish brown, variegated with cream color, the latter tint deepening here and there into tan, or paling into white; variegations closer about lower neck; throat pale buff; remainder of neck (all around) and most of head marked with dusky streaks and dots upon a buff or pale-brown ground; top of head darker; iridescence of speculum very imperfect,

often scarcely discernible.   Under parts of plumage pale buff, or
dull white, with obscure spots or freckles.   Bill uniformly dusky.
Legs and feet bluish gray.

Length twenty-two and a half to twenty-three inches: extent nearly that of drake.

Young (both sexes).   Closely resembling adult female.

"Northern Hemisphere.   In North America breeds from the
northern parts of the United States northward, and migrates
south to Cuba and Panama" (A. O. U. Check List).

I have heard no local name applied from Calais, Me., to Bath,
though between these points I have interviewed many duckers.
The bird is certainly not common enough here to require often
a name of any kind, and it may be added that nowhere upon
our coast is the species so numerous as in the interior.   From
Bath to the State of Connecticut the name **GRAY DUCK** (see
Nos. 6, 9) is usually given it (I find that I have thus lumped the
matter in notes relating to this portion of the coast), but no
other name has troubled me so much as this one.   It can be appropriately applied to many species, and is too comprehensive,
too adaptable a title to remain as unwaveringly attached to a
single species as do duck-names usually.   It is very liable to be
brought into play when a grayish duck of any kind is shot that
the gunners are unfamiliar with.   Though I have met the name
in a large majority of the places visited, I have only been able
to record its exact local use in a comparatively few instances.
Referring to its general application in New England to the
present species, Mr. Brewster writes (Bull. Nutt. Orn. Club, July,
1883): "Much confusion has been caused by the assumption
that the Gray Duck (*i. e.*, Gadwall) of the books is the same with
the 'Gray Duck' of New England gunners and sportsmen."   I
have heard this name popularly applied to the species now in
hand, on the Niagara (see No. 6); in Connecticut, at Essex, Milford, and Stratford; at Bellport, Long Island; Washington, D. C.;
and Alexandria, Va.; and very commonly in these localities, as
elsewhere, to designate only the females, and the males in gray
attire.   Giraud says, in his Birds of Long Island, 1844, referring

3*

to this fowl: "The young and females are mistaken by many persons for a distinct species, which they call Gray Duck."

Some of the duckers of Seaford (Hempstead), Long Island, include full-plumaged drake under above title, while others distinguish the latter as **PIED GRAY DUCK** (see note to No. 23, concerning use of "pied" on Long Island).

In Philadelphia, Baltimore, and St. Augustine, these young birds and females are also called Gray Duck, and in the latter locality **GRAY WIDGEON** (see No. 9); latter name likewise used more or less at Essex, Conn., where the species goes by the name of **SEA WIDGEON** as well.

Also known (including or not the *gray* birds) on the Niagara; about Lake St. Clair; in Massachusetts at Salem, North Scituate, North Plymouth, and West Barnstable, as **PIN-TAIL** (see No. 31), and at Salem and West Barnstable as **SPLIT-TAIL.** At Buzzard's Bay, Mass.; in Connecticut, at Essex and Stratford; in New Jersey, at Barnegat, Tuckerton, Pleasantville (Atlantic Co.), and Atlantic City; at Baltimore, Washington, Alexandria, Norfolk, and at Morehead, N. C., **SPRIG-TAIL**; this being sometimes shortened to **SPRIG.** At Chicago, **SPIKE-TAIL,** and less commonly **PIKE-TAIL**; at Milford, Conn., **PICKET-TAIL**; this being probably the original form of a Long Island name, which I find spelled "Picketail" in my note book, and which Giraud gives as "Picitail" in the index to his Birds of Long Island. Several old duckers conversed with at Shinnecock Bay, Moriches, Bellport, etc., consider this a corruption of *peaked-tail,* but I imagine they are a little off the track.

To the older gunners about Milford, this is the **PHEASANT DUCK** or **PHEASANT**; and similar names by which the species has been known are **SEA PHEASANT** and **WATER PHEASANT.** For other water-fowl to which the name "pheasant" is applied, see Nos. 20, 21, 22.

In New Jersey, at Manasquan (Monmouth Co.), **SMEE**; at Tuckerton, **SMEES**; while others at Tuckerton refer to it as **SMETHE.** Though these are doubtless forms of one and the same name, I have thought best to make no choice between them, but to give the three equal prominence. The species has been so termed, it appears, for a very long time. "Most of us," said a venerable hunter, "call it Sprig-tail, but I suppose its real name is Smees." Josselyn, in his Voyages to New England, published 1674, mentions "Smethes" among other of our birds, but to what species he referred we can only guess. It is interesting also to recall the fact that the Smew or White Nun, *Mergus albellus* (no longer included in our fauna), has been called both "Smee" and "Smeath."

At Pleasantville and Atlantic City we hear LONG-NECK, and at Charleston and Savannah SPRIG-TAILED WIDGEON; while in Charleston markets and to some of the local gunners it is the WIDGEON simply. (See our Widgeon of the books, No. 8; also Nos. 9, 12, 17, 31.) At St. Augustine the full-feathered drake is the KITE-TAILED WIDGEON.

Other aliases gathered from various sources, but that I do not remember having heard in common use, are WINTER DUCK (Nuttall—See No. 25); CRACKER (Fleming's British Animals); SPREET-TAIL, PILE-START (both in Giraud's Birds of Long Island); PIGEON-TAIL (Herbert's Field Sports); SHARP-TAIL (Hallock's Gazetteer, and Long's American Wild Fowl Shooting); SPINDLE-TAIL (Water Birds of North America). And Rev. Chas. Swainson, in Provincial Names of British Birds, 1885, gives "LADY BIRD (Dublin Bay)," and "HARLAN (Wexford)," recording also a name previously mentioned, as follows: "Sea Pheasant (Hants; Dorset)."

No. 14.

Spatula clypeata.

Adult male. Head and upper neck of a very dark greenish tone, with purple reflections; lower neck and breast white; belly and flanks rich chestnut brown; front part of wings conspicuously blue, of light shade, but vivid; back of this blue, a green wing-mark, or speculum, bordered with white and black; feath-

No. 14. Adult male.

ers striped with white, sweeping backward from inner region of wings; back dusky brown; rump, and above and below tail, black with greenish gloss; at either side of tail a white patch.

Audubon wrote: "We have no duck in the United States whose plumage is more changeable than that of the male of this beautiful species." The species can nevertheless be quickly recognized in any plumage by its broad, soft, and yielding bill, twice as wide at its rounded end as at its base, with fringe of fine tooth-like processes exposing themselves on either side.

I have never had the good-fortune to kill a drake in above splendid attire, and have taken my description mainly from three stuffed specimens shot near Savannah, Ga. The bills of these are black, but the colors of bills and legs before death I can only determine through the testimony of others. According to several ornithological works, the bills (accompanying this or similar plumage) are black, or nearly so; the legs reddish orange, or vermilion.

No. 14. Female.

I have shot many of this species in the late fall and early winter in Southern and Western States, but they were either drakes in imperfect plumage, or birds in female apparel. This latter dress is a simple mixture of warm brown and light buff, iine and streaky on head and upper neck : the throat plain buff. Lower neck (all around) and general upper plumage brown, the feathers edged with light buff which pales to white on broader feathers growing from shoulder regions, or inner region of wings; the feathers covering sides of body marked in nearly like manner. Forward portion of wings brownish, with light markings near front edge (no blue); speculum greenish, bordered before and very narrowly behind with white, and often very dull, with little or no lustre. Lower surface of body varying from buff to nearly pure white, mottled about vent and beneath tail with warm brown.

Length about twenty inches; extent thirty inches or more.

Bill, as I have commonly observed it on freshly killed birds, but as I have never seen it described: upper division (or upper mandible) olive brown, with bright orange edge, the surface dotted with black as though fly-specked; lower division bright orange;* these colors changing rapidly after death.

Many imperfectly plumaged drakes that I have seen have dark head and neck, finely speckled with white; snowy white and dark markings about breast and back; front of wing blue to greater or less degree, and dull brownish leather color on belly.

I have always found this duck fine eating. Audubon says: "The sportsman who is a judge will never pass a Shoveller to shoot a Canvas-back."

Range, "Northern Hemisphere. In North America breeding from Alaska to Texas" (A. O. U. Check List).

SHOVELLER: BLUE-WING SHOVELLER (Catesby's Nat. Hist. of Carolina, etc.): **RED-BREASTED SHOVELLER** (Pennant's British Zoölogy).

Along the coast from New Brunswick to Connecticut this species is too rare to bear a well-established name among gunners. It is known at Lake St. Clair; the Detroit River; Chicago; Long Island; in New Jersey at Red Bank (Monmouth Co.), Barnegat, Atlantic City, and Sommers Point; in Maryland at Havre de Grace and Baltimore; in Virginia at Alex-

---

* Catesby, 1731, describes bill as "reddish brown, spotted with black" (his specimen being in brown plumage, with front of wing blue); and in Water Birds of North America (Baird, Brewer, and Ridgway), bill colors of adult female are thus described: "bill brown, mandible orange;" but no mention is made of the black dots.

andria and Norfolk; at Morehead, N. C., and Savannah, Ga., as SPOON-BILL * (see No. 31).

At Tuckerton, N. J., and Crisfield, Md., it is the SHOVEL-BILL, and in Putnam County, Illinois, the BROADY.

The name BROAD-BILL, given in Yarrell's British Birds, Coues's Key, etc., though eminently appropriate, seems to have been very thoroughly taken up in our country by other species.

Another name at Norfolk, and one which has rapidly grown into favor, is BUTLER DUCK, the bird being so called because of its spoon-like bill, and with reference to a well-known general in the civil war. J. W. Long also records this name in his descriptions of wildfowl shooting in the West.

Another odd title, of much less recent origin, encountered at Morehead, N. C., is COW-FROG. Though no one attempts to give a reason for the term, the oldest inhabitants tell of hearing it in use from early childhood.

Though known at Savannah, as previously stated, as the Spoon-bill, I have heard it oftener referred to there, and at St. Augustine, as SPOON-BILLED WIDGEON; and it is commonly called in the markets, and by the market-gunners of Savannah, the SPOON-BILLED TEAL. This termination "teal," though a peculiarly marketable one, is not applied in this case from mercenary motives alone, as many of the resident sportsmen as well as market gunners believe in two varieties of Spoon-bill; the Spoon-billed Widgeon being the larger, and having "darker bill and legs."

The only time I remember to have heard the name Shoveller in actual use among gunners (and this, according to scholarly usage, is its correct name) was at Baltimore. The bird is known however as the MUD-SHOVELLER at Sanford, Fla.

In Lawson's New Voyage to Carolina, 1709, we read about the SWADDLE-BILL as follows: "A sort of an ash-colored duck, which have an extraordinary broad bill, and are good meat;

---

* Our Roseate Spoonbill, allied to the herons, and known to ornithologists by the weird and double-barrelled title *Ajaja ajaja*, will not, it is hoped, get mixed in the mind of any one with the duck kind.

they are not common as the others are." As Pennant remarks in his Arctic Zoölogy, 1785, referring to the above (but without naming Lawson): "We must therefore join it, for the present, to this species."

In Swainson's Provincial Names of British Birds, 1885, we find SHOVELARD (Norfolk); MAIDEN DUCK (Wexford); SHEL-DRAKE and WHINYARD (Waterford); "whinyard" being "the name for a knife like the Shoveller's bill in shape." Mr. Swainson also states that the name Whinyard is given in Wexford to the European Pochard *Fuligula ferina*.

No. 15.

Aythya vallisneria.

Adult Male. Head and greater part of neck brownish red
or mahogany color; top of head and about bill of deeper tint, or
blackish; reddish tone extending farther down neck than in
species No. 16 (sometimes confounded with No. 15), and approaching less nearly a true red; remainder of neck, fore part

No. 15 Adult Male.

of body, and rump blackish brown; wings principally gray.
Back, shoulder-feathers, sides, and about vent white, delicately
dotted and lined in wavy pattern with dusky gray ("wrapt in
pencilled snow"); front of wings with wavy markings in similar
fashion. Under parts of body, not previously described, pure

white. Bill high at base, greenish black throughout; its length from corner of mouth two and a half to two and three-quarter inches, and greatest width about thirteen sixteenths of an inch. Legs bluish gray.

Female. Head, neck, and fore part of body dull brownish buff or brownish tan; wings nearly plain grayish brown; upper parts of body, with sides, and rump all around grayish brown minutely sprinkled with wavy dull white. Belly white, tinged here and there with yellowish and grayish tints. Bill and legs as in male.

No. 15. Female.

Length twenty-one to twenty-two and three-quarter inches: extent thirty-three and a half to thirty-six inches or a little more.

Another way of distinguishing it from No. 16 is by the rather flat manner in which the forehead continues the upper line of the bill; the forehead of No. 16 being more arched and intellectual-looking.

In many waters of the West, this bird, whose range is supposed to include the greater part of the country,* is found in

* "Breeds from the northern tier of states northward, in the Rocky Moun-

goodly numbers, but it is not a common species anywhere near our Eastern coast north of Delaware; and in New England it is rare.

A friend who has spent much time at Norfolk, Va., informs me that a majority of the Norfolk epicures consider this bird better eating when it first arrives from the North than it is at any other time. This is antagonistic with the popular belief that the "wild celery" of the Chesapeake region does so much to improve the bird's flavor.

Wilson, who first described this species (scientifically), tells us (1814) of its being called **CANVAS-BACK** on the Susquehanna. **WHITE-BACK** on the Potomac, and **SHELDRAKE** (see Nos. 20, 21, 22) on the James. Jefferson, in his Notes on Virginia (ed. 1788), mentions "Sheldrach, or Canvas-back;" and the name "White-back" is still a familiar one to duckers on the Potomac, at least to those about Washington and Alexandria.

Wilson tells us also of a wheat-laden vessel wrecked near Great Egg Harbor, N. J., and how the floating grain attracted vast numbers of these birds, which, being unknown to the local gunners, were denominated " sea-duck " simply; and Ord adds, in his reprint of Wilson, that in the neighborhood of Philadelphia hunters were in the habit of supplying the market with this duck, under the name of " Red-head," or " Red-neck " (see No. 16), and that "their ignorance of its being the true Canvas-back was cunningly fostered by our neighbors of the Chesapeake, who boldly asserted that only their waters were favored with this species." Audubon speaks of Southern epicures sending to Baltimore for Canvas-back, not knowing that they could be obtained near home. " I well remember," he writes, "that on my pointing out to a friend, now alas, dead, several dozens of these birds in the market of Savannah, he would scarcely believe that I was not mistaken, and assured me that they were looked upon as poor, dry and fishy." But now, this " over-rated and

tains further south, and in upper California; winters in the United States, and southward to Guatemala."—*Coues.*

generally under-done bird," as Dr. Coues nicely puts it, is recognized as the "Canvas-back" almost everywhere. It may be added that in ducking parlance the abbreviation **CAN** is sometimes used; I should not emphasize this fact had not the abbreviation crept into print occasionally as a distinct name, without apology or explanation.

Known to many gunners about Morehead, N C., and on New River, Onslow Co., same state. as **BULL-NECK** (see No. 31), and in last-named locality, as **RED-HEADED BULL-NECK**.

No. 16.

Aythya americana.

Adult male. Head and upper neck mahogany colored; head large with full puffy feathering; remainder of neck, fore breast, and around on extreme forward part of back continuously black or blackish. Plumage of back and sides finely zigzagged with dull white and slaty black; the lower back and tail grayish

No. 16. Adult male.

brown; and immediately about tail blackish. Wings principally two shades of bluish gray, their under surfaces grayish and white. Under surface of body white, shading darker with brownish gray towards tail. Bill pale blue (in life) with black end; length of bill, measured along edge from corner of mouth,

4

two to two and a quarter inches; its greatest width thirteen sixteenths to fourteen sixteenths of an inch. Legs bluish gray.

Female. Bill similar to drake's, but darker in color; head and upper neck drab or grayish brown; immediately about bill and throat lighter grayish buff. Lower regions of neck, upper parts of body, and the sides brown and slaty brown; edges of the feathers paler, the pale edging more noticeable about lower neck region and sides. No zigzag markings anywhere (or with barest suggestion of them). Wing much as in male. Under

No. 16.   Female.

parts white, shading darker and brownish gray behind. Legs as in male.

Length twenty to twenty-one inches: extent about thirty-three inches.

Range, North America in general, breeding from Maine and California northward.

RED-HEAD, or **RED-HEADED DUCK:** very generally known as such in the books, and by gunners. It has been also called the **POCHARD** from its resemblance to European Pochard (with which it was at one time considered identical), and more correctly the **AMERICAN POCHARD.**

At Seaford (Hempstead), L. I., it is the **RED-HEADED BROAD-BILL.** Upon the coast north of Long Island this species, though occasionally killed, is certainly far from a familiar sight to gunners.

From Pamlico Sound to South Carolina commonly known as the **RED-HEADED RAFT-DUCK.**

In King's Sportsman and Naturalist in Canada, 1866, called **GRAY-BACK** (see No. 17); and in Schoolcraft, 1820, and Tanner's Narrative, 1830, **FALL DUCK.**

Another name, too interesting to be omitted, is found in Avifauna Columbiana (Coues and Prentiss, 1883), *i. e.,* **WASHINGTON CANVAS-BACK,** accompanied by the following remarks : " One of the commonest market ducks, passing about half the time for Canvas-back, and equally available for promoting Congressional legislation."

No. 17.

Aythya marila nearctica.

Adult male. Head, neck, fore part of breast, and front of
back black; the gloss about head chiefly greenish. Lower part
of back, the long wing-feathers, and tail mostly brownish black.
Wing-mark, or speculum, white. Middle of back, sides of body,
and flanks beautifully pencilled with black and white zigzag

No 17. Adult Male.

lines, the same extending on to wings, but less conspicuously.
Under surface of body principally white, though pencilled like
flanks on lower belly, and blackish beneath tail. Bill broad, and
of light bluish lead color with black nail at end. Legs and feet
gray, with blackish webs. Eyes yellow.

Female. Dark brown in those parts which are black in male, though with front of head immediately around base of bill white. The zigzag markings much less distinct. Bill less blue, more dusky, and often with dull olive tinge. In other respects resembling drake.

No. 17. Female.

"Total length about eighteen to twenty inches; extent twenty-nine and a half to thirty-five inches." *

Range, North America in general; "breeding far north" (A. O. U. Check List).

Concerning the edibleness of this and following species, No. 18, doctors disagree, as in many similar cases; quality of flesh depending so much upon feeding-ground. My own experience has led me to agree with Wilson, that "their flesh is not of the most delicate kind."

AMERICAN SCAUP DUCK: GREATER SCAUP DUCK. "Scaup,"

---

* Having omitted to take measurements myself, I quote Water Birds of North America (Baird, Brewer, and Ridgway).

according to Willoughby, is a term applied to broken shell-fish; and Yarrell, treating of British Birds, says: "Beds of oysters and mussels are in the north called 'oyster-scaup' and 'mussel-scaup,' and from feeding on these shell-covered banks the bird has obtained the name."

It is impossible to separate clearly the names of this from those of the following species, No. 18, the two being enough alike to travel very commonly under one and the same name.

Along the coast from New Brunswick, until approaching Long Island Sound, duckers do not usually remark a difference between them; and I had better state here, once for all, that the following names, which are not specially remarked upon as applied to this, the greater scaup, alone, may be regarded as belonging to both species.

Known in Maine at Jonesport, Frenchman's Bay, Ash Point (near Rockland), Portland, and Pine Point, and in Massachusetts at Salem, Barnstable, Fairhaven, and Falmouth, as **BLUE BILL.*** This is the popular appellation in the West also. I have met it in common use on the Niagara and Illinois Rivers, at Chicago, and about Lake St. Clair; and Mr. J. P. Leach, of Rushville, Ill., writes me concerning this and No. 18 in his part of the country, that they are " almost invariably known as ' blue-bills;' the terms ' broad-bill,' ' scaup,' ' black-head,' etc., rarely being used except by men from the East."

We hear " blue-bill " also (among other names) in New Jersey at Pleasantville (Atlantic Co.) and Cape May City; and infrequently used at Jacksonville, Fla.

This, the larger scaup, is distinguished in the vicinity of Detroit and Lake St. Clair, as **LAKE BLUE-BILL,** and this name is recorded as "local" in the Revised List of Birds of Central New York, 1879 (Rathbun, Fowler, and others).

Again at Falmouth, Mass., and to native duckers at Newport, R. I., **WIDGEON** (see our Widgeon of ornithologists, No. 8, also Nos. 9, 12, 13, 31): in Boston markets **BLUE-BILLED WIDG-**

---

* Given at Machias Port, Me., to Ruddy Duck, No. 31.

EON: at North Plymouth, same state, AMERICAN WIDGEON*
(to some at least); and Mr. F. C. Browne gives TROOP-FOWL in
his list of "gunners' names" at Plymouth Bay (Forest and
Stream, Nov. 9, 1876).

. Another title at Chicago is GRAY-BACK (see No. 16), and
certain gunners about Detroit prefer BLACK-NECK to the more
common Western term "blue-bill."

In Connecticut at Stonington, mouth of Connecticut River,
Stony Creek, and Stratford, BROAD-BILL (this being monopolized
at Bath, Me., and Newport, R. I., by Ruddy Duck, No. 31; see
also No. 14). I find latter name in like use in New Jersey at
Barnegat, Tuckerton, Pleasantville (Atlantic Co.), and Cape May
City, and in Virginia at Richmond.

The greater scaup is distinguished on Long Island at Shin-
necock Bay and Moriches as BAY BROAD-BILL, and again at
Shinnecock Bay as DEEP-WATER BROAD-BILL; at Bellport as
WINTER BROAD-BILL; and at Manasquan, N. J., as SALT-WATER
BROAD-BILL. Another name once common about Shinnecock
Bay, but now seldom heard, is MUSSEL-DUCK.

Again at Pleasantville, N. J., and at Crisfield, Md., FLOCK-
DUCK; Crisfield duckers frequently, however, distinguish the
greater scaup as GREEN-HEAD (see No. 6, to which this name
is usually applied).

At the mouth of the Susquehanna, very commonly on the
Chesapeake, by some at Cape May C. H., at Eastville and
Cobb's Island, Va., and at Charleston, S. C., BLACK-HEAD; the
greater being particularized on the Chesapeake as BAY BLACK-
HEAD. I have heard the term "black-head" as far south as
St. Augustine, though SEA-DUCK and RAFT-DUCK are names
better understood by St. Augustine natives; the latter name
being equally popular at Savannah and Jacksonville.

At Washington, D. C., and Alexandria, Va., BAY-SHUFFLER.
At Morehead, N. C., and Wilmington, same state, BLACK-

---

* The American Widgeon of the books, No. 8, being locally known as
"California Widgeon." Scaups are not common enough in the neighborhood of Plymouth to require often a name of any kind, and No. 8 is rare.

HEADED RAFT-DUCK; this being applied only to greater scaup in first-named locality. By others at Wilmington, **BULL-NECK** (see Nos. 15, 31).

It may be added that the words "big" and "large" as prefixes to "blue-bill," "broad-bill," etc., frequently distinguish this from the following scaup.

No. 18.

Aythya affinis.

In appearance, excepting size, like No. 17; gloss about head, however (when discernible), having more of a purplish than greenish cast.

"Length fifteen and a half to seventeen inches; extent under thirty inches." *

Range, "North America in general; breeding chiefly north of the United States" (A. O. U. Check List).

See No. 17 for names shared in common with this species, but not repeated here.

This, the **LESSER SCAUP DUCK,** though often found in company with preceding species, is certainly more partial to inland water, or rivers, creeks, and ponds. I find it locally designated as follows:

In the vicinity of Lake St. Clair, **RIVER BLUE-BILL;** at Chicago, **LITTLE BLUE-BILL** (adjectives "little" and "small" being naturally more or less used in many localities by those who recognize Nos. 17 and 18 as distinct); in Revised List of Birds of Central New York, 1879, **MARSH BLUE-BILL** (given as "local"); at Pleasantville, Atlantic Co., New Jersey, **MUD BLUE-BILL.**

At Stratford, Connecticut, and Seaford (Hempstead), L. I., **RIVER BROAD-BILL;** at Shinnecock Bay, L. I., and Tucker-

---

* Having quoted measurements of greater scaup, No. 17, I prefer to quote these also, this time from Dr. Coues's Key, 1884. By comparing quotations, and taking into account that the plumages are practically alike, it will be seen that under certain conditions the two ducks are not easily separated.

ton, N. J., **CREEK BROAD-BILL**; and Giraud in his Birds of Long Island, 1844, mentions this name as "well-known to the bay gunners."

Again, at Bellport, L. I., **MUD BROAD-BILL**; and in New Jersey, at Barnegat, **FRESH-WATER BROAD-BILL**; at Cape May C. H., **GOSHEN BROAD-BILL** (the cove at Goshen, same county, being a favorite resort of this smaller species); in Abbott's catalogue of New Jersey Birds, 1868, **POND BROAD-BILL**.

Many duckers of the Chesapeake know it as **CREEK BLACK-HEAD**; and in Virginia at Eastville and Cobb's Island it is the **FLOCK DUCK**; and Audubon speaks of its being known in Kentucky as **FLOCKING-FOWL**.

At Washington, D. C., and Alexandria, Va., **RIVER SHUFFLER**; and in Newberne and Morehead, N. C., **SHUFFLER**, simply; the latter name being never intentionally applied in the vicinity of Morehead, at least, to greater scaup.

No. 19.

Aythya collaris.

Adult male. Bill dark slate color with blackish tip and bands of pale blue, a narrow band at base, and broader one near end. A brownish red or mahogany colored ring around neck. Head and neck above ring black, with slight iridescence, and white chin-mark. Neck below ring, fore-breast, and plumage of upper parts blackish. Wings slaty brown; the wing-mark, or

No. 19.   Adult Male.

speculum, bluish gray. Sides of body finely waved with white and blackish lines. Under parts white, with dusky markings towards rear, and black, or nearly black, beneath tail. Legs gray.

Female. No ring around neck, and no wavy lines anywhere. Bill much less plainly marked. Dark parts of plumage brownish; the front of head about base of bill whitish, this face-marking purer white on chin, and mixed more or less with brown on forehead. Wings marked with bluish gray as in

No. 19. Female.

male. Lower surface of body grayish white, freckled with brown or grayish brown, and becoming more uniformly brown behind.

Length sixteen to eighteen inches; extent, say from twenty-five to twenty-seven inches. A much less numerous species than either No. 17 or 18.

Range, North America in general; breeding north of the United States. Though found along the coast, oftener met with in the interior.

RING-NECKED DUCK: RING-NECKED SCAUP: RING-NECKED BLACK-HEAD: TUFTED-DUCK of Wilson.

In the vicinity of Lake St. Clair **MARSH BLUE-BILL** (see No. 18), a name given also by Mr. Seton in his Birds of Western Manitoba (Auk, April, 1886).

At Chicago **RING-BILL**; and Audubon speaks of hearing the latter name in Kentucky.

In Putnam Co., Ill., **BLACK-JACK**; this being heard also at Chicago, though less commonly; and Mr. J. P. Leach, of Rushville, Ill., writes that this name is "generally applied along the Illinois River."

In Porter's Spirit of the Times, Oct. 25th, 1856, it is reported as travelling in the vicinity of Cincinnati, under "the euphonious but unmeaning" title of **BUNTY**; and Mr. Long, in his American Wild Fowl Shooting, 1874, gives **GOLDEN-EYES** as a "very common" name in the West (see No. 23, to which "Golden-eye" is usually applied).

Giraud writes (referring to this duck) in his Birds of Long Island, 1844: "By our gunners generally it is considered a hybrid, and familiar to them by the name of **BASTARD BROAD-BILL**."

At the mouth of the Susquehanna commonly known as **RING-BILLED BLACK-HEAD**, though many of the local gunners regard the female as a distinct species, and term it **CREEK RED-HEAD**, because of its resemblance to female No. 16.

At Newberne, N. C., and Wilmington, same state, **RING-BILLED SHUFFLER**; and I am told by two well-informed gunners, viz., Alonzo Nye, of Chatham, Mass., and William Flint, of Lyme, Conn., of its being known to certain South Carolina duckers as the **MOON-BILL**.

No. 20.

Merganser americanus.

Bill rather cylindrical in shape, hooked at point, and furnished with very positive teeth, entirely unlike what is understood as a *duck* bill; see outline drawings under No. 21.

Adult Male. Head and upper neck dark green, or blackish with green gloss; a slight, scarcely noticeable crest. Eyes and bill red; the latter, however, having its ridge, and "nail" at

No. 20. Adult Male.

end black. Lower neck all around, and under parts of body white, more or less deeply tinged (in life) with buff. Sides of rump faintly waved with gray. Tail, and lower part of back, gray; remainder of back black. Wings white and black, as indicated in picture. Legs red.

Female. Head and upper neck, including crest, reddish

brown; the crest more fully developed than in male. Throat, lower neck, and marking on wing (as shown in picture), white. Upper parts, generally, and sides slaty gray. Under surface of body cream colored (in life). Legs, bill, and eyes much as in male, but less bright.

No. 20. Female.

Measurements about as follows: *Male:* length twenty-six to twenty-seven inches; extent thirty-six inches. *Female:* length twenty-four inches; extent thirty-five inches.

Though this bird is found in both salt and fresh water, it belongs much more to the inland than following merganser No. 21. It is a thoroughly cold-weather creature, remaining on lakes, ponds, etc., as long as a single "breathing-hole" is left in the ice, and, having been forced to depart, it returns at the very first show of open water, ascertaining the fact immediately in an altogether marvellous manner.

Range, North America in general; breeding from northern border of United States, northward.

GOOSANDER (a name commonly regarded as from *goose* and

gander; "a goosey-goosey-gander" sort of name): **MERGAN-
SER** (diver-goose, Latin *mergus* and *anser*): **AMERICAN MER-
GANSER** (distinguishing it from European species, *Merganser
merganser*): **DUN-DIVER** of Pennant (relating to dun color of
female plumage; see No. 31): **SPARLING FOWL** of Latham
(a good name for this fish-eating fowl, "sparling" being an
old English name for the smelt): **BUFF-BREASTED SHEL-
DRAKE**:* **BUFF-BREASTED MERGANSER: AMERICAN SHEL-
DRAKE.**

In Maine at Eastport and Milbridge, and in Massachusetts
at Pigeon Cove and Salem, **SHELDRAKE,**† indiscriminately
with No. 21; the present species monopolizing said title at
Rowley, Mass., on the Niagara River, in the neighborhood of
Chicago, at Hennepin and Snachwine, Ill., and Morehead, N. C.

At Ellsworth, Me., and in Massachusetts at North Plymouth,
Buzzard's Bay, and West Barnstable, **POND SHELDRAKE** (see
No. 22).

At Bath and Kennebunk, Me., and Portsmouth, N. H.,
WINTER SHELDRAKE; at Pine Point, Me., **GREAT LAKE SHEL-
DRAKE;** and at West Barnstable, Mass., **SWAMP SHELDRAKE**
(as well as Pond Sheldrake. See No. 22 for name Swamp Sheldrake as used on Long Island).

At Falmouth, Mass., and in New Jersey at Barnegat and
Tuckerton, **FRESH-WATER SHELDRAKE;** and in latter state, at
Pleasantville (Atlantic Co.), Atlantic City, and Somers Point,
RIVER SHELDRAKE; and at Pleasantville again, **NORTH CARO-
LINA SHELDRAKE.** It may add interest to note in this con-
nection, as well as farther on, that the term "Carolina" is also

---

* The name Sheldrake is probably from provincial English *sheld*, meaning
variegated or pied, and *drake.*

Yarrell says of the Old World Sheldrake: "I have found the stomach of
this species filled with very minute bivalve and univalve mollusca only, as
though they had sought no other food; a predilection which may have given
rise to the name of Shell-drake; or it may be so called because it is parti-
colored; and the term Shield-drake may have had its origin in the frequent
use made of this bird in Heraldry."

† See odd use of this name under No. 15.

employed at Pleasantville for the Hudsonian Godwit, No. 61;
the latter bird being locally known as the Carolina Willet.

On Buzzard's Bay, from New Bedford to Barney's Joy Point,
BREAKHORN; at Stonington, Conn., **BRACKET SHELDRAKE**, or
BRACKET simply. The meaning of these two names, Breakhorn
and Bracket, I cannot give, and I think that the latter has been
introduced in Stonington since I learned to shoot there thirty-
five years ago.

On Long Island at Moriches, **WEASER SHELDRAKE**; at Bellport and Seaford (Hempstead), **WEASER**. The term "sheldrake," that is to say, being more commonly omitted in latter
localities. I have given the spelling of Giraud, who refers to
this name in his Birds of Long Island, 1844. Another form of
it, or I should say a name that immediately suggests the other,
is heard at Shinnecock Bay (designating same species), viz.:
TWEEZER. I can hardly believe that this last is the original
form, though the bird's beak is easily likened to a pair of
tweezers. My idea is that early settlers on the Island associated
our "fresh-water sheldrake" with the German river, and got to
calling it in consequence the *Weser* sheldrake.

Though, as previously mentioned, the name "sheldrake" at
Milbridge, Me., commonly includes next species, No. 21, some of
the older gunners there distinguish No. 20 as the **PHEASANT**; the
latter distinction being general at Machias Port and Jonesport,
same state; and we read in Wilson, Vol. VIII., 1814, of this
species (No. 20) being "called by some the **WATER-PHEASANT.**"
(For other water-fowl to which the name "pheasant" is attached
see Nos. 13, 21, 22.)

At Milford and Stratford, Conn., **VELVET-BREAST**; and to
some Atlantic City gunners, **MOROCCO-HEAD**.

Other names by which our three mergansers (Nos. 20, 21, 22)
have been more or less loosely known, are **FISH-DUCK, FISHING-
DUCK, FISHERMAN**, and **SAW-BILL**. The names Fisherman and
Fishing-duck are, however, monopolized in the neighborhood of
Morehead, N. C., by No. 21.

5

No. 21.

Merganser serrator.

Adult male. Head, long hairy crest, and a little of neck black, with greenish gloss; the neck beneath this, white; lower neck and fore-breast a rather light chestnut brown, speckled streakily with black. Front of back and the wings black and

No. 21. Adult Male.

white, as in picture; this black connected with that of head by black line on back of neck; remainder of back and sides of body a wavy pattern of narrow white and black lines. Under parts white. Bill and legs red, the former dusky on top.

Female. Head and upper part of neck principally reddish

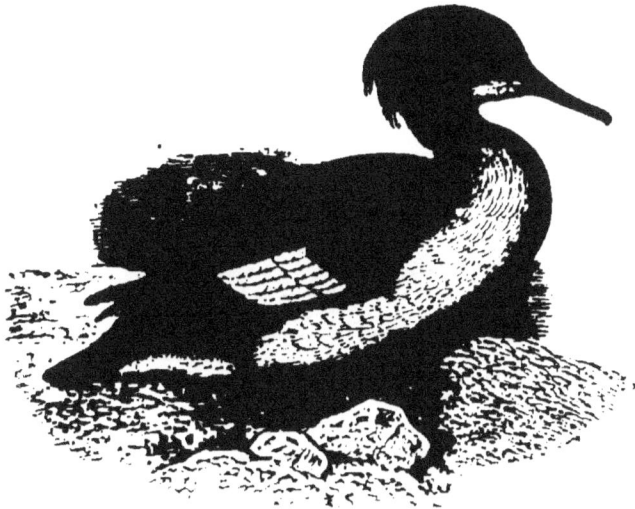

No. 21. Female.

brown (of duller tint, commonly, than in female No. 20), the upper part of head darker, of grayish tone; a streak of this gray running back along by eye from bill, the throat paling to white, remainder of neck, back, sides, and tail brownish gray, with pale edgings to the feathers; wing chiefly dark brown, marked with white as in picture. Under parts white. Bill and legs of duller tint than in adult male.

Length twenty-two to twenty-four inches; extent thirty-two to thirty-four inches.

A very common water-fowl. Range, as given in A. O. U. Check List: "Northern portions of northern hemisphere; south, in winter, throughout the United States."

The female of this species is easily confused with that of No. 20. Note difference in line of feathering at base of bill, and position of nostrils, on following page.

Our mergansers, all three of them, are much better eating than commonly supposed; though, as is the case with numerous species, they are less desirable in some localities than in others. Once, while at Lane's, on Shinnecock Bay, L. I., I had the fun of watching a gentleman, who regarded fowl of this sort with holy horror, ravenously devour a bird of the present species.

He was charmingly ignorant, of course, of what he ate, but
when informed a short time after, he not only succeeded in re-
taining his food, but confessed like a man that Shinnecock sheldrakes, at least, were a success.

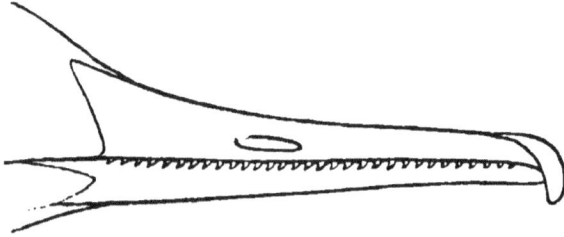

Bill of No. 20.

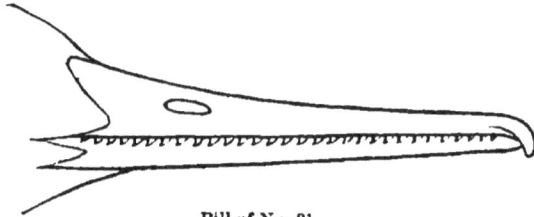

Bill of No. 21.

RED-BREASTED MERGANSER. RED-BREASTED GOOSANDER
of Edwards's Natural History of Birds,Vol. II., 1747.

Very generally known as **SHELDRAKE** from Eastport, Me.,
to— I can only say, lower waters of Chesapeake, as I have
no note of hearing it used for the species farther south than
Eastville, Northampton Co., Va. (See No. 20, a bird sometimes
confounded with this; and also odd use of "Sheldrake" under
No. 15.) Certain names more especially distinguishing this from
"Sheldrakes" Nos. 20 and 22, and other exceptions to above-
mentioned general use of the name, are noted as follows:

At Bath, Me., **SPRING-SHELDRAKE**; at Rowley, Mass., **SEA-
ROBIN**, or **ROBIN** simply; at Stonington, Conn., though the
name Sheldrake is more or less used for both sexes, many gunners distinguish the female as **SHELDUCK**, and this latter form
is common for both sexes in New Jersey at Pleasantville (Atlantic Co.), Cape May C. H., and Cape May City.

At Essex, Conn., **LONG ISLAND SHELDRAKE**; at Barnegat

and Tuckerton, N. J., **SALT-WATER SHELDRAKE**; and Giraud writes, Birds of Long Island, 1844, "called by our gunners **PIED SHELDRAKE**" (see No. 23 for use of "pied" on Long Island).

At Crisfield, Md., **PHEASANT**. (For other water-fowl to which the word "pheasant" is attached, see Nos. 13, 20, 22.)

At Morehead, N. C., **FISHERMAN** and **FISHING-DUCK**. These last two names, though used at Morehead for this bird only, are sometimes loosely applied to the three mergansers (Nos. 20, 21, 22). We also hear **FISH-DUCK** and **SAW-BILL** thus indiscriminately applied; and William F. Davis, of the Thimble Islands, Conn., tells of hearing the name **GAR-BILL** used for mergansers in general, "by visiting sportsmen" from parts unknown.

Captain Bob Petty, of Bellport, L. I., informs me that this, the Red-breasted Merganser, is known "to all the gunners about Mobile" (Ala.) as the **SEA BEC-SCIE** (this being an English-French combination, meaning sea "saw-bill").

In a Notice of the Ducks and Shooting of the Chesapeake, by Dr. J. T. Sharpless, Cabinet of Nat. Hist., Vol. III., 1833, the present species is referred to as **HAIRY-CROWN**, a name reminding us of that similar one, Hairy-head, belonging to Hooded Merganser, No. 22.

De Kay, in New York Zoölogy, 1844, mentions "Whistler" among other names, as given to this species in New York State. He elsewhere records the title as applied in same state to the Hooded Merganser. I do not feel like giving special emphasis to these applications of a term so commonly used, then as now, for the Golden-eye, No. 23.

We find the following in Rev. Charles Swainson's Provincial Names of British Birds, 1885: **SAWNEB** (Aberdeen): **SAWBILL WIDGEON** (Galway): **HERALD** (Shetland Isles): **HERALD DUCK** (Forfar and Shetland Isles): **HARLE** or **HARLE DUCK** (Orkney Isles): **EARL DUCK** (East Lothian): **LAND HARLAN** (Wexford): **BARDRAKE** (Down), "from the brown and ash colored streak on the rump;" this name being mentioned elsewhere by the author as applied in Ireland to *Tadorna cornuta*, the common Sheldrake of the Old Country: **SCALE DUCK** (Strangford

Lough): **GRAY DIVER** (Islay) "applied to the female:" **POP-PING WIDGEON** (Drogheda Bay); and Mr. Swainson tells of this latter name being used in same locality ("Drogheda Bay") for the European Golden-eye, "as it pops down and up so suddenly."

No. 22.

Lophodytes cucullatus.

Adult male. Bill nearly black, shorter than in preceding mergansers, and differing slightly in other respects, yet, nevertheless, a "saw-bill." Crest black in front, and white behind with black bordering. Head, neck, and much of upper plumage

Merganser. No. 22. Adult Male. Little

black, with some brown, and occasional greenish reflections; loose black feathers striped with white, growing from elbow region, and seeming (while wings are closed) to belong rather to the plumage of the lower back; wing-mark, or speculum, white, with black bars. Sides of body cinnamon brown, finely waved with dark lines; breast and belly white, the white of breast and black of upper parts sweeping into each other, and forming crescent-like bars in front of wings. Legs yellow brown. Eyes yellow.

Female. A little smaller than male. Upper parts brownish, with no pure black; crest rusty brown, with no white, and

No. 22. Female.

smaller than drake's; front of breast grayish, and without the crescent bars; throat and under parts white or nearly so. Bill blackish above and orange below (similar in shape to that of male).

Young. Practically like adult female.

Length seventeen to eighteen inches; extent about twenty-five inches.

These birds are very partial to fresh water, and when near the sea are met with usually in small rivers, creeks, and ponds. They are peculiarly sportive and agile, and easily decoyed by anything resembling a duck. The beautiful fan-like crest is lifted or lowered at will.

Range, North America in general; breeding here and there throughout the United States and northward.

HOODED MERGANSER: HOODED SHELDRAKE: ROUND-CREST-
ED DUCK (Catesby's Nat. Hist. Carolina, Fla., etc., 1731): FAN-
CRESTED DUCK (Barton's Fragments Nat. Hist., Penn., 1799).

At Bath, Me., PICKAXE SHELDRAKE (the bill being the pointed
end of the pickaxe, I suppose; the crest, its wide transverse edge):
known also at Bath, to some of the gunners, and at Essex, Conn.
as POND SHELDRAKE (see No. 20): and Mr. Everett Smith states
in his Birds of Maine,* that it is " locally known as the LITTLE
SHELDRAKE."

At Stonington, Conn., WOOD SHELDRAKE; at Essex, same
state, SUMMER SHELDRAKE. Neither this name Summer Sheldrake, nor that of Pond Sheldrake is often required here, as the
bird is but infrequently found, and it may be added that this is
not a common species along our coast north of New Jersey,
though met with sometimes in fair numbers.

On Long Island at Shinnecock Bay, Moriches, and Bellport,
SWAMP SHELDRAKE (see No. 20).

On the Niagara River, Lake St. Clair, and about Chicago,
LITTLE SAW-BILL and FISH-DUCK; the latter name being common also in Putnam Co., Ill. (See No. 21 for this last name, and
Fisherman, Fishing-duck, Saw-bill, and Gar-bill, as sometimes
indiscriminately applied to mergansers in general : a loose style
of expression, however, that belongs more to "sportsmen" and
the like than to "gunners.")

It is worthy of note that in the neighborhood of Niagara
Falls the book-name, Hooded Merganser, is met with in common
use. Just think of it! a live gunner with that name on his
lips.

In Connecticut at Milford and Stratford, SAW-BILL DIVER.
I am here reminded of how easily names get twisted. I have
seen this one conspicuously printed "Swan-bill Diver," and an
old gunner at Stratford always refers to the bird as " Saw-mill
Diver;" the last being not so bad, as the bird is so frequently
encountered in and about mill-ponds.

On Long Island at Seaford (Hempstead), SAW-BILL simply;

---

* *Forest and Stream*, 1882–83.

a name distinguishing it here from other mergansers, but, as elsewhere stated, sometimes loosely employed to designate the three (Nos. 20, 21, 22) collectively.

In Abbott's catalogue of New Jersey birds, 1868, we read of the present species being "generally known inland" as **POND SAW-BILL.**

At Detroit, **SPIKE-BILL.** Nowhere in western localities mentioned have I heard the name "sheldrake" applied to it.

At Newport, R. I., **SMEW.** The Hooded Merganser is about the size of the true Smew, *Mergus albellus*, and the drake of the latter species, when his crest is erected, looks considerably like our bird ; very much as our bird might look in a state of partial albinism. The Smew proper is no longer included in our fauna, and it is doubtful if it ever should have been. Though Wilson tells us that it was "frequently observed" in his time "in the ponds of New England," etc.

At Manasquan, N. J., **WATER-PHEASANT;** at Morehead, N. C., **PHEASANT DUCK,** and more commonly **PHEASANT*** simply. Lawson writes, in his New Voyage to Carolina, 1709 : "The water-pheasant (very improperly called so) are a waterfowl of the duck kind, having a topping of pretty feathers which sets them out." (For other water-fowl to which "pheasant" is attached, see Nos. 13, 20, 21.)

In New Jersey at Barnegat, Tuckerton, Pleasantville (Atlantic Co.), Atlantic City, and Somers Point, **COCK-ROBIN,** and less commonly **COCK-ROBIN DUCK;** at Somers Point, Cape May C. H., and Cape May City, and at Eastville, Va., Wilmington, N. C., and St. Augustine, Fla., **HAIRY-HEAD.**

At Crisfield, Md. (east shore of Chesapeake), **SNOWL;** a name as weird as some of those in Alice's Wonderland, and the only one by which the bird is known, so far, at least, as I could discover in 1885.

To the darkies of Charleston, S. C., and its vicinity, **MOSS-**

---

* The Ruffed Grouse, No. 41, generally known by this name in the South, is not met with in this section, and when referred to is termed "Mountain Pheasant."

HEAD. The colored women often use a large bunch of "Florida Moss," *Tillandsia usneoides*, as a cushion for the heavy loads they carry upon their heads, and I am inclined to believe that " Moss-head" was suggested by this practice, rather than by any direct resemblance to moss in the bird's crest.

I find also in my memorandum-book the name **TOW-HEAD** for this species, but, unfortunately, with no note of locality accompanying it. I remember distinctly, however, that the name was heard in one of our Southern States.

Another name (than that of "Hairy-head") commonly heard among the " crackers" of St. Augustine is **TADPOLE**; the bird ' having been thought particularly fond of polliwogs, I suppose.

While examining specimens in the Smithsonian (Washington, D. C.), I was surprised to find the name **WOOD-DUCK** (see No. 12) printed on this bird's label. But Mr. Ridgway told me that he had heard " Wood-duck," and also **TREE-DUCK** (again see No. 12) commonly applied to this species, in lower or more southern portions of the Wabash valley, Ill. and Ind. The application of " Wood-duck" to a " Saw-bill," though a little shocking at first, is natural enough, of course, as the Merganser breeds in woods, nesting in the hollow of a tree like the " Wood-duck " of people generally; and Mr. George A. Boardman, of Calais, Me., once witnessed a lively and long-continued fight between a bird of the latter species and a Hooded Merganser for the possession of a hole in a tree to which both laid claim.

I have previously quoted Captain Petty, for the Red-breasted Merganser. The captain adds that the present species is known to all about Mobile, as **BEC-SCIE**; this (the French for " Saw-bill ") distinguishing it from the Sea Bec-scie, No. 21.

Glaucionetta clangula americana.

Adult male. Head and upper neck black (or of very deep tone), richly glossed with green; a roundish spot of white between bill and eye. Remainder of neck, with lower parts of body, pure white excepting a few brownish gray mottlings about vent and sides of belly. Back, wings, and tail practi-

No. 23. Adult Male.

cally black (here and there blackish-brown), with white markings as shown or sufficiently indicated in picture. Bill black or nearly so; eyes bright yellow; legs and toes yellow or orange, with dusky webs.

Female. Considerably smaller than male; head plain brown; neck in front and at sides white faintly touched with gray; behind brownish gray. Lower part of neck, with fore-breast, and upper parts generally, gray, the feathers pale at edges; wings darker, with white markings as indicated in picture. Under parts white, the color of upper plumage continued down about

No. 23. Female.

the legs and behind them. Eyes as in male. Bill dull yellowish, or yellowish olive, shaded unevenly with blackish brown. Legs and toes dull yellow with dusky shading, the webs chiefly black.

Length seventeen to twenty inches; extent twenty-seven to thirty-one inches.

A duck more or less common in winter throughout the country, making its appearance, as Giraud says, "about the same time that a majority of its tribe are compelled to quit the 'great nursery' at the North for our more temperate climate."

AMERICAN GOLDEN-EYE: COMMON GOLDEN-EYE. As the first name marks this bird as different from European variety, so the second distinguishes it from Barrow's Golden-eye,* a species of our own which is *not* "common," to Eastern gunners at least.

MORRILLON (Arctic Zoölogy, 1785): **GARROT**, another Old World title early applied to our bird: **CONJURING-DUCK: SPIRIT-DUCK.** Richardson, 1831, speaks of these last two names as given in the fur countries to both this species and No. 24, because of their instantaneous disappearance "at the flash of a gun or the twang of a bow."

"Sometimes called by our gunners the **BRASS-EYED WHIST-LER**" (Nuttall's Water Birds, Boston, 1834). **BRASS-EYE**, mentioned by DeKay, Zoölogy of New York, 1844.

From Eastport, Me., to Falmouth, Mass., on the Niagara River, at Chicago, along the Connecticut coast, and at Shinnecock Bay, L. I., **WHISTLER.**

At Milford, Conn., and Shinnecock, the adult drake, though recognized by all as of the same species with the rest, is commonly referred to as the "pied Whistler."†

At Niagara Falls, Chicago, Newport, R. I., and Alexandria, Va., **WHISTLE-WING;** at Cape May C. H., N. J., **WHISTLE DUCK;** and we find this latter form in Beesley's Birds of Cape May, 1857.

Another and very pretty name, heard at Lyme, Conn., but almost exclusively among the old people, is **MERRY-WING.** A disagreement, however, exists concerning its use, whether it rightfully belongs to this fowl or the following, No. 24. Having obtained equally reliable testimony on both sides I record the name in both lists.

---

* The Barrow's Golden-eye, or Rocky Mountain Garrot, is very similar in general appearance to the present species, but the adult drake has the white patch between the bill and eye crescent-shaped, and the species are in other ways distinguishable.

† The word "pied" is peculiarly popular on Long Island, where the gunners prefix it to local names to designate the "full dressed" male of any species whose plumage *is* pied or showily variegated, and when I asked an old ducker if he did not think the present species particularly handsome, he said, "Yes, the *pied* ones are very handsome."

At Plymouth, Mass., though "Whistler" is the more common appellation, we occasionally hear that of GOLDEN-EYE, * and this latter name is the common one at Detroit, and we meet with it (among other names) at Chicago.

At Seaford (Hempstead), L. I., GREAT-HEAD;† in New Jersey at Barnegat, Tuckerton, Pleasantville (Atlantic Co.), Atlantic City, and Somers Point, CUB-HEAD; at Cape May C. H., COB-HEAD, the last name being monopolized, however, by the young birds, which are regarded as a species distinct from the "Whistle-ducks." At Havre de Grace, Md., BULL-HEAD; at Morehead, N. C., IRON-HEAD. The name Cob-head is again heard at Cape May City, where the species is also very generally known as CUR; a name that may have come from likening the bird's note to that of a dog.‡ But whatever the origin, this rather contemptuous title certainly has the charm of brevity, and is, in this respect at least, preferable to " *Glaucionetta clangula americana.*"

At Pleasantville (before mentioned), JINGLER; at Baltimore and on the Patapsco River, WHIFFLER; at Crisfield (Somerset Co.), Md., KING DIVER.

---

* See No. 19 for "Golden-*eye.*"

† Giraud writes, Birds of Long Island, 1844, "by some it is called Great Head."

‡ Since writing the above, I have found that in portions of Great Britain the name "Curre" is given to the Golden-eye *C. clangula;* and Swainson says, in his Provincial Names of British Birds, that this is " from the bird's croaking cry."

Charitonetta albeola.

Adult male. Head very dark or blackish, richly glossed with purple and green, a field of white from the eye backward. Back black, fading to pale gray near tail; wings and shoulder feathers principally white and black, as in picture; the long wing-feathers and the tail gray. Neck, continuously with under

No. 24. Adult Male.

parts of plumage, white, the latter shaded a little about legs and tail with pale brown. Bill leaden blue, the nail and about base dusky. Legs and feet very light flesh color with lavender tinge.

Female. Considerably smaller than male, and without the full fluffy feathering of head. Plumage of upper parts brown,

shading on fore-breast and sides to gray or grayish brown; spot
on side of head and wing-mark white. Lower parts white,
though with some dusky shading about the legs and back of

No. 24. Female.

them. Bill more dusky than in male. Legs bluish gray with
lavender tinge, the webs dusky.

Length twelve and three quarters to fifteen inches; extent
twenty-two to twenty-five inches.

This is a common species, visiting most parts of the country
during winter, and the full-dressed drake is one of our most
beautiful birds.

BUFFLE-HEAD, or **BUFFEL'S HEAD DUCK** as Catesby gives
it (Nat. Hist. Carolina, Florida, etc., 1731): **BUFFLE DUCK**:
BUFFALO-HEADED DUCK: LITTLE BROWN DUCK, the female
being described under this latter title in another part of Catesby's work: **SPIRIT,** or **SPIRIT-DUCK:** Edwards, in Nat. Hist. Birds,
Part II., 1747, describes the drake as **LITTLE BLACK AND WHITE
DUCK,** and speaks of its being known to Newfoundland fishermen as " Spirit:" **CONJURING-DUCK,** see Conjuring and Spirit
Duck, No. 23.

Very generally known from Eastport, Me., to Falmouth,
Mass., as **DIPPER**; * though at certain points along this coast it

---

* This and other names mentioned farther along are considerably mixed

is too rare to bear a name of any kind. Having been told at Kennebunk, Me. (1885), that a very handsome but strange duck had recently been killed, I walked a long distance out of my way to see it, and was considerably disappointed to find the *rara avis* nothing more wonderful than a male of the present species. Again, while at Provincetown, Mass. (same year), I was called to pronounce upon another cock Dipper, as the bird was unknown to the local gunners (see "Dipper" of Provincetown, No. 31).

I will note here that the Water Ouzel, *Cinclus mexicanus*, also bears the name of "Dipper" in books and elsewhere, but there is little chance of confusion arising therefrom, the Water Ouzel being about the size of a blue-bird, and belonging to the far West.

The Buffle-head is again the "Dipper" on the Connecticut coast, and continues to be so recognized, very generally, as far as the southern part of North Carolina.

At Bath, Me., and North Scituate, Mass., **ROBIN-DIPPER**; at Buzzard's Bay, Mass., **DAPPER** (see No. 31). Mr. Browne, in his list of gunners' names, at Plymouth Bay,* gives both "Dipper" and **DOPPER** (see No. 31).

DIE-DIPPER (see foot-note to name "Dipper," page 81): **MARRIONETTE**: these two names being mentioned by Audubon, that of Marrionette belonging to the state of Louisiana.

"Devil-diver" and "Hell-diver" have also appeared in print once or twice as aliases of this bird, but I do not feel like emphasizing the fact; I have never heard either of them used by a gunner for any bird but a grebe, and I think they have probably been credited to the present species inadvertently.

---

with those of the Ruddy Duck, No. 31, and with the Grebes, particularly the Pied-billed Grebe, *Podilymbus podiceps*, that lively little nuisance, familiar to us all, under one or more of the following titles: Hen-bill, Hen-bill Diver, Hell-diver, Devil-diver, Water-witch, Dab-chick, Dob-chick, Dop-chick, Dip-chick, Die-dapper, Die-dipper, Dipper. I do not mean, however, that the same name is applied in any one locality to more than a single species.

* *Forest and Stream*, Nov. 9, 1876.

At Niagara Falls, Lake St. Clair, Chicago, Snachwine (Putnam Co.), Ill., Washington, D. C., Charleston, S. C., and Savannah, Ga., **BUTTER-BALL** (see No. 31); and in last-named city, **BUTTER-DUCK** (see No. 31). The name Dipper, which is much more commonly used for this species at Washington, is given to the Pied-billed Grebe in the western localities mentioned.

Bartram, in Travels through North and South Carolina, etc., 1791, speaks of the Buffle-head being "called **BUTTERBACK.**" Wilson writes: "Usually known by the name of the **BUTTER-BOX**, or Butter-ball," and Nuttall gives "Butter-box" as used "in Pennsylvania and New Jersey."

In New Jersey at Pleasantville (Atlantic Co.), Cape May C. H., and Cape May City, **DIVER**; to some at Norfolk, Va., and Currituck region, **WOOL-HEAD**; at Wilmington, N. C., **SCOTCH-DUCK**, **SCOTCHMAN**, **SCOTCH-DIPPER**, and **SCOTCH-TEAL**; the latter name being a favorite with hucksters, "Teal" being always in demand.

No. 25.

Clangula hyemalis.

Adult male in winter. Markings as in picture, and practically brownish black and white; the patch about eye dove gray, and the dark patch next this tint (on side of neck) blackish above and brown below; the blackish breast tinged more or less deeply with chocolate; the feathers which sweep acutely backward from shoulders grayish white; the long and slender pair of central

No. 25. Adult Male in winter.

tail-feathers having an outward and slightly upward curve. Bill, from head half-way to tip, and nail at end black, the remainder light rose pink. Legs and toes light bluish gray, with joints dusky and webs blackish.

Adult male in summer. In the spring, before this bird leaves us for the North, its summer dress is more or less fully assumed. A drake shot April 12th, at Stony Creek, Conn., whose nuptial plumage was complete, or very nearly so, was dressed as follows: Head, neck, breast, and upper parts generally, deep chocolate brown intensified to pure black here and there; an irregularly outlined space of light mouse gray on each side of head, from bill backward to include the eye; the eye itself irregularly bordered with white; a patch of bright tan-colored feathers with black centres just behind neck on front of back, and feathers similarly colored sweeping from shoulder regions along the sides of the back. Under parts and sides, from the dark breast backward, white, this white tinged with pearly gray upon the sides, and meeting abruptly the deep tint of breast and upper plumage. Bill, legs, and long tail-feathers as in winter male, though light color of bill better described, perhaps, in this case as salmon pink * (a delicate tint that darkened in a few hours to reddish purple). Weight two pounds.

No. 25. Female in winter.

Adult female in winter. Head and most of neck white,

---

* As this light color is not given as I describe it, in any of the books accessible to me at this writing, I will state that, so far as both bill and legs are concerned, my notes were made (in this case and in that of the winter drake) within ten minutes after the bird was shot.

with top of head and patch on each cheek blackish; also blackish or dusky at chin, this chin-marking continued narrowly and
imperfectly downward to lower neck; the lower neck and extreme fore part of body chiefly dusky gray or brownish gray,
this deepening in tint or blackening rather suddenly immediately
below the white of the neck, producing a collar-like marking in
some cases.   Upper parts of body including wings chiefly dark
or blackish brown, variegated at neck, sides of breast, and on scap-
ulars (shoulder-feathers) with gray and reddish brown.   Lower
surface and sides of body broadly white.   Bill olive gray with
dusky shading (the olive tint not always noticeable).   Legs and
feet as in male.

Young female.   Similar to adult female just described, but
with upper parts more uniformly dark.

Young male.   When this youngster first comes to us from
the North he is in general appearance much like the winter
female (tail, color of bill, and all); but he is a bigger bird, and
while passing from this stage to that of adult drake his variations are altogether too complicated for description.

Having omitted to note with sufficient care summer plumage

No. 25.   Female in summer.

of female, I will quote Audubon, who was familiar with the species in its breeding-ground: " The head is dark grayish brown with a patch of grayish white surrounding the eye, but not extending to the bill; there is a larger patch of the same color on the side of the neck, the hind part of which is similar to the head, the fore part grayish brown, the feathers broadly margined with whitish. All the upper parts are of a dark grayish brown, the two lateral tail-feathers edged with white; the lower parts white, the feathers under the wings slightly tinged with gray."

Measurements about as follows: *Male:* with tail fully developed, length twenty-three and a half inches; extent thirty inches. *Female:* length fifteen to sixteen inches; extent twenty-eight inches.

This bird is not popularly regarded as very desirable for table use, though it is relished by many gunners, and I have myself found it as good as some of the Canvas-backs which I have killed on prairie ponds and tried to eat. Its flight is peculiarly swift, irregular, and very swallow-like; it is a crafty and enduring diver, a lover of cold weather, and eminently a sea-duck, though found on certain inland waters as indicated in list of local names.

Its range is given in Ridgway's Manual, 1887, as "Northern portions of northern hemisphere; in America, south in winter, to nearly across the United States."

LONG-TAILED DUCK, of early as well as late authors: LONG-TAILED HARELD * (Selby's Illust. Brit. Orn.): SWALLOW-TAILED DUCK, so termed at Hudson's Bay (Fauna Boreali-Americana, 1831): NOISY DUCK, because of its " reiterated cries " (Audubon): HOUND, a name applied in Newfoundland (the musical gabble of a flock being likened to the cry of hounds).

Known all along the New England coast as OLD SQUAW, the full-feathered drake being sometimes distinguished, as at West Barnstable and Fairhaven, Mass., Stonington and Essex, Conn., as OLD INJUN.

---

* Hareld is the same as *Haveld*, an Icelandic name for the species.

On Long Island we find the name Old Squaw dividing honors with that of **OLD WIFE,** the latter continuing in more or less general use to the sea-coast of Maryland. South of this, to Eastville, Va. (I have no note of meeting with the species farther south), and on Chesapeake Bay, it is the **SOUTH SOUTHERLY,** frequently pronounced Sou' Southerly, and a corruption of this, viz., **SOU' SOUTHERLAND,** is also common. The names Old Squaw and Old Wife are very rarely heard on this latter piece of coast.

At Crisfield, Md. (east shore of Chesapeake), **SOUTHERLY,** and at Eastville, Va., **SOUTHERLAND.**

Not one of the three old duckers conversed with at Seaford (Hempstead), L. I. (1884), had heard any of these " southerly " names, and at Crisfield, Md. (same year), I could find no one who had heard " Old Squaw." I remember that while learning to shoot, at Stonington, Conn., some thirty-five years ago, I was more familiar with the name South Southerly and its elongated form, **SOUTH-SOUTH SOUTHERLY,** than with any other.

Wilson says (Vol. VIII., 1814), " This duck is very generally known along the shores of the Chesapeake Bay by the name of South Southerly, from the singularity of its cry, something imitative of the sound of these words, and, also, that when very clamorous they are supposed to betoken a southerly wind; on the coast of New Jersey they are usually called Old Wives."

I am told that in Stonington, Conn., the words " John Connolly " were popularly used, about fifty years ago, in imitation of this bird's gabble, and they can be so repeated as to produce a better imitation, I think, than the words now in use at Stony Creek, same state, viz., " Uncle Huldy," and " my Aunt Huldy."

In New Jersey, at Pleasantville (Atlantic Co.), and Somers Point, **OLD MOLLY;** at Atlantic City and Somers Point, **OLD GRANNY,** and **GRANNY** simply; at Cape May City, **MOMMY;** the drake being distinguished at Pleasantville as **OLD BILLY.**

On the Niagara River, and about Lake St. Clair, **COWEEN;** and McIlwraith writes, in his Birds of Ontario, 1886, " Vast numbers of ' cowheens' (as these birds are called here) spend the winter in Lake Ontario." Known also to French Canadians and

others at Detroit and St. Clair Flats, and to the people of Kennebunk, Me., as **COCKAWEE**; this (differently spelled) being referred to in Fauna Boreali-Americana as follows : " The peculiar
cry of this duck is celebrated in the songs of the Canadian voyageurs, by the epithet of *caccàwee*," and Mr. William Brewster
speaks of "cock-a-wee" as everywhere applied to the species on
the Gulf of St. Lawrence (1883).

Two other odd names met with among old New England
gunners are **SCOLDENORE**, at Portsmouth, N. H., and **QUANDY**,
at North Scituate and Plymouth, Mass. We hear at Plymouth
also, **SCOLDER**, a term much more easily understood.

To some at Lake St. Clair and Chicago this is the **WINTER-
DUCK** (see No. 13), while others at Chicago are more familiar
with the New England title Old Squaw.

The following names are found in Swainson's Provincial
Names of British Birds, 1885 : **SHARP-TAILED DUCK: SWALLOW-
TAILED SHELDRAKE: CALOO,** or **CALAW** (Orkney, Shetland Isles):
DARCALL: COAL AND CANDLE-LIGHT (Orkney Isles): **COL-
CANDLE-WICK** (Fife): **COLDIE** (Forfar): **MEALY BIRD** (Norfolk),
"the young are so called:" **NORTHERN HARELD** (Aberdeen).

Histrionicus histrionicus.

Adult male. Prevailing color bluish slate with more or less purplish tinge, becoming brownish beneath; the head and neck purplish black; plumage fantastically slashed and spotted with white, as picture shows better than any written description can; the white markings intensified here and there by edging of pure black; at either side of crown (or top of head) a stripe of ma-

No. 26. Adult Male.

hogany red, and a broad patch of the same color on either flank. Bill yellowish olive with tip lighter. Legs and toes light bluish gray, with blackish webs.

Female. A very different looking fowl, nearly all grayish brown; side of head marked with dull white, and white mixing with the lower plumage and producing a dull freckled belly. Bill and legs dull bluish gray.

No. 90. Female.

Young. Practically like adult female.

Length sixteen and a half to seventeen and a half inches; extent twenty-four and a half to twenty-seven inches : bill much narrower towards end ; length on top one inch or a trifle more (from feathers to tip).

The range of this species includes the northern part of North America. It is found, in winter, as far south as Massachusetts, and on very rare occasions a little farther south. In many places along the Maine coast it is a common and well-known bird.

HARLEQUIN DUCK of authors generally : PAINTED DUCK and MOUNTAIN DUCK of "Hudson's Bay residents," according to Fauna Boreali-Americana, 1831 : known about Mud and Seal Islands, Yarmouth Co., Nova Scotia, as ROCK DUCK, so says Rev. J. H. Langille, in Our Birds in their Haunts, 1884.

Along the coast from New Brunswick to Salem, Mass., LORD AND LADY; farther south than this the species is rare, and I have no note of hearing gunners name it. Known also as SQUEALER at Machias Port, Me., and as LORD simply, at Jonesport, same state. Edwards, in Natural History of Birds, Part II., 1747, speaks of this "the Dusky and Spotted Duck" being sent from Newfoundland, where the "fishers call it the Lord."

No. 27.

Somateria dresseri.

Adult male. Chiefly black and white as shown in picture, the black having a brownish cast here and there, and the white tinged more or less upon the breast with buff yellow; black of head glossed with purplish blue and divided behind by white; on hinder part of head and along the lower edge of the black

No. 27. Adult Male.

marking a wash of sea green. Bill extending back peculiarly upon the forehead, this extension (in life) of a yielding, leathery character, divided or forked behind into two broad branches, or lobes, with roundish ends, these lobes sweeping to right and left upon sides of forehead; measurements (taken from a single speci-

men) as follows : from tip to extreme back line of lobes, three
and one eighth inches ; from tip to point of feathering where
lobes meet, or where the leathery extension begins to divide, two
and a quarter inches ; width of each lobe five eighths of an inch.

Female. Plumage chiefly a speckled and barred mixture of
light brown or tan color, and black, the bar-like markings more
decided along the sides, and the markings of lower parts indistinct
or blurred. Bill similar in general style to that of male, but
branching more narrowly upon the forehead ; its measurements

No. 27. Female.

(taken from a single specimen) as follows: from tip to extreme
back line of lobes a little less than three inches; from tip to
central point of feathering where the lobes meet two and a
quarter inches ; width of each lobe five sixteenths of an inch.

Having failed to note colors of bills and legs in perfectly fresh
specimens, I will quote Audubon. *Male:* Bill pale grayish yellow, the unguis (nail at end) lighter, the soft tumid part pale
flesh color ; feet dingy light green, the webs dusky. *Female:*
Bill pale grayish green ; feet as in the male.

Length twenty-four to twenty-six inches; extent thirty-nine to forty-two inches.

AMERICAN EIDER: COMMON EIDER: very generally known along the coast from New Brunswick to Rhode Island as SEA DUCK, or SEA DUCK AND DRAKE; at Barnstable, Mass., SHOAL DUCK and ISLES OF SHOALS DUCK; the latter name being likewise heard at New Bedford, same state, and in Connecticut at Stonington and Stony Creek: known also at New Bedford and Stony Creek as WAMP (this being of Indian origin, probably; *wompi*, white).

Eiders are Northern birds, and are seldom seen on the Connecticut coast, though they congregate every winter in large flocks in Muskegat Channel, at the west end of Nantucket, and sometimes, it is said, wander as far south as the Delaware.

Giraud, in Birds of Long Island, 1844, speaks of the species being called SQUAM DUCK in Maine, and De Kay in Zoölogy of New York, 1844, of its being known on Long Island as BLACK AND WHITE COOT and BIG SEA DUCK. The latter author states also that it is called Squaw Duck on the Maine coast, but I regard this as simply a misprint of name previously mentioned by Giraud. (Though these books bear same date, Giraud's was first published.)

The common Eider of Europe, *Somateria mollissima*, is known as Dunter, Dunter Goose, Dunter Duck, and Cuthbert Duck or Saint Cuthbert's Duck, among other names; I add these because until a few years ago ornithologists regarded the two birds as one and the same. With the exception of a rather slight difference in the shape of the bill * there is little or no difference between them, and the difference between the bills of the females of the two species is in some cases very difficult to detect.

The superiority of the down of the eider every one is more or less acquainted with, and the flesh is said to be very good under certain conditions, but I have never tried it. Audu-

---

* In *mollissima* the elongated encroachments of bill upon forehead are narrower, and run back straighter, and terminate more acutely.

bon tells of their being sold at Boston, in the winter of 1832
(when they were far more common than now), " at from fifty to
seventy-five cents the pair;" and he adds " they are much sought
after by epicures."

Note.—The three following species, Nos. 28, 29, and 30, known as "coots" (see Nos. 31, 32, 33, 37) or "sea coots," "scoters," etc., cannot be regarded as general favorites for the table, and we can fully understand the reason when we attempt the mastication of one of mature years. Latham writes concerning this kind of fowl (Synopsis, 1785): "The flesh tastes fishy to an extreme, and from this cause is allowed by the Roman Catholics to be eaten on fast days, and in Lent; and indeed, to say the truth, must be a sufficient mortification."

I am forced to omit many local names heard for these birds, finding it impossible to obtain a satisfactory vote among the duckers and fishermen as to which species they belong to. The three are in many ways similar, and the females differ enough from the old cocks to be often classed as distinct varieties.

In the markets of Washington, D. C., the name "booby" (see No. 31) is indiscriminately applied to fowl of this genus (not often killed, however, so far up the river), and they are referred to collectively, and facetiously, at Pleasantville (Atlantic Co.), N. J., as "iron pots," or "old iron pots."

No. 28.

Oidemia deglandi.

Adult male. Plumage black, with conspicuous band of white on wing, and small patch of white on side of head sweeping backward with upward curve from lower part of eye; eye pearl white with small black centre or pupil, the white of the patch below narrowly continued around edge of eye. Bill with abruptly rising knob at base; much encroached upon by feathers; immediately at base black, this black spreading forward over the knob and continued along the edge to nail at end; sides of bill purplish red changing to orange near base; nail also orange, and from nail to the black between the nostrils white, or pearl white; the middle of the bill, in other words from nail to knob, being broadly white; lower mandible (lower division of bill) black with broad patch of orange (including nail) at end, this patch paling to white at back edge. The legs may be briefly described as red, with black joints and webs, but the two sides of legs and toes differ considerably in color, the outside surfaces

No. 28. Adult Male.

being dull purplish pink, the inner bright carmine red pervaded more or less with orange.

Female. Sooty brown, lighter and more gray below, and to some extent whitish on sides of head, this white appearing very differently upon different specimens, often as speckles between the eye and the bill, and as a condensed blotch behind the eye.

No. 28. Female

Wing marked with white as in male. Bill swollen, but without
the abrupt knob, and uniformly blackish; the eyes also very
dark. Legs and feet dull flesh color heavily shaded with black,
and with webs black.

Young male. Resembling female, but darker or more blackish about head and neck; also noticeably darker on lower surface
of body, and showing, generally before the beginning of winter,
pinkish tinge on sides of bill (where the old drake is purplish
red), and having brighter and a little more reddish legs.

In any plumage this species is instantly distinguished from
Nos. 29 and 30 by the white wing-mark alone.

Measurements (highest and lowest of eighteen freshly killed
birds): length nineteen and five eighths to twenty-two and five
eighths inches; extent thirty-three and seven eighths to forty
and three sixteenth inches.

Range, as given in A. O. U. Check List, 1886, "Northern
North America, breeding in Labrador and the fur countries;
south in winter to the Middle States, Southern Illinois, and
Southern California." I cannot remember killing one of these
birds on the eastern coast south of the above limit, though they
are to be found farther south, doubtless. Dr. Coues says (Key,
1884): "North America at large," etc.

WHITE-WINGED SCOTER: WHITE-WINGED SURF DUCK: VELVET SCOTER: VELVET DUCK.

From New Brunswick to Chesapeake region (in localities far
too numerous to mention) **WHITE-WINGED COOT** or **WHITE-WING**.

In Massachusetts at Pigeon Cove and North Scituate, **BLACK
WHITE-WING** for adult drake, and **GRAY WHITE-WING** for female and young; some gunners believing that these two plumages represent separate species.

Again in Massachusetts at New Bedford and Fairhaven, and
southward along the shore to Newport, R. I., the name **MAY
WHITE-WING** or **GREAT MAY WHITE-WING** is applied to certain
full-feathered birds, locally regarded as a distinct and larger
variety, to be met with only between the 10th and 20th of May

"flying west nor'west." Many duckers tell me that these larger birds are seldom or never seen to alight, and that they almost always appear late in the afternoon, and are to be seen passing over in immense flocks until hidden by the night. This supposed "variety" is also called, though less frequently, the **EASTERN WHITE-WING**, on Buzzard's Bay from New Bedford to Barney's Joy Point. As I myself have never witnessed this May migration, the above account is all that I can give concerning it.

At Portsmouth, N. H., and at Rowley and Salem, Mass., **PIED-WINGED COOT**; in Connecticut at Milford (to the older gunners) **BELL-TONGUE COOT**, at Stratford **UNCLE-SAM COOT**; on Long Island at Bellport **BULL COOT,*** at Moriches **BRANT COOT**; to some at Portsmouth, N. H., **SEA BRANT**; in the neighborhood of Niagara Falls **BLACK DUCK** (see Nos. 7, 29); at Crisfield, Md., **ASSEMBLYMAN** (though known as White-wing also), the species being commonly referred to, singly or collectively, as *'Semblymen.*

Mentioned in A Notice of the Ducks and Shooting of the Chesapeake, by Dr. Sharpless (Cab. Nat. Hist., Vol. III., 1833), as "Velvet, or **CHANNEL DUCK**."

The "Lake Huron Scoter" described and figured by Herbert ("Frank Forester") in appendix to Field Sports, Vol II., was of this species, and the author's testimony concerning its flesh is amusing, when we think how disgusted he would have been to have known that his scoter was simply the bird previously described in same volume as "coarse," "fishy," "tough," and "worthless." Having gotten hold of a young bird, however, and excited with the belief that he had added a new species to our fauna, he gushed as follows: "Not only as fat and as juicy, but as delicate, as tender, as lusciously melting in the mouth, as any

---

* Since writing the above, I have heard this name at Stony Creek, Conn., and it strikes me as peculiarly appropriate for these thick-necked, big-headed, heavily built drakes. The heaviest of five males, shot December 8th, weighed four and a half pounds; and gunners tell me that *May* White-wings sometimes weigh considerably more.

Gunpowder River Canvas I ever had the fortune to taste." Poor Herbert! though you were about right in these latter remarks, you were not the man to have written them, had you known that you were comparing your aristocratic " Canvas " to such a common and despised fowl as this.

No. 29.

Oidemia perspicillata.

/

Adult male. Plumage black; a patch of white on forehead, another white patch on nape, also a scarcely noticeable touch of white immediately below the eye upon the lid. Bill lifting high over nostrils, its upper sides bulging outward abruptly at base and free from feather-encroachment, but the black feathering of

No. 29. Adult Male.

forehead continued along the top of the bill to nostrils. At its sides this very conspicuous beak is pure white from the base halfway to tip, with squarish patch of black (as shown in picture), this courtplaster-like patch being separated from the feathering on top of the bill by a line of orange, and from the black of the plumage behind by a narrower line of carmine red; top or ridge of bill (including nostrils) deep carmine red, this changing to

7*

bright orange and spreading over the sides of the bill in front; nail at end yellow; lower mandible, or lower division of bill, white shaded with orange in front, the nail yellow (like its fellow above). Eyes white or pearl-white with black pupil (the female having dark eyes as in the case of species last described). Legs and feet red, with joints blackish and webs pure black; that is to say, they may be briefly so described, but (as in the case of preceding species, No. 28), the two sides of legs and toes differ considerably, the outer surfaces being carmine red, the inner orange with more or less carmine red tinge.

No. 29.   Female.

Female. Sooty brown with lower surface of body gray; no patch on forehead or nape, but more or less whitish on sides of head in blotches. Bill but slightly swollen about base, not feathered so far forward on top, and uniformly dusky, or gray very fully shaded with black. Legs and feet dull brownish yellow with dusky shading, the webs black.

Young male. Much like female, but showing, often before winter, some pinkish tinge at sides of bill in front.

Measurements (highest and lowest of five freshly killed birds): length seventeen and a half to nineteen and a half inches; extent thirty and a half to thirty-two and a half inches. Weight of an adult male killed in December, two pounds and three and a quarter ounces.

Range: chiefly coasts of North America, but also found on inland waters; breeding far north, and moving south in winter to the Carolinas, Ohio and Kansas rivers, Lower California, and even to the Island of Jamaica.

SURF SCOTER: SURF DUCK: BLACK DUCK of Pennant, 1785 (see Nos. 7, 28): in Edwards's Natural History of Birds, Part III., 1750, it is "the great Black Duck from Hudson's Bay."

In Maine at Eastport, Millbridge, Bois Bupert Island, Frenchman's Bay, and Portland, **HORSE-HEAD COOT**, or **HORSE-HEAD**; to some at Eastport, **BALD-PATE** (see No. 8); at Machiasport and Jonesport, **SKUNK-BILL**; at Portsmouth, N. H., in Massachusetts at Pigeon Cove, Cohasset, North Scituate, North Plymouth, Barnstable, Chatham, and Falmouth, at Stony Creek, Conn., and on Long Island at Shinnecock Bay and Moriches, **SKUNK-HEAD** (the name Skunk-*bill* being, doubtless, a perversion); at Essex, Conn., **SKUNK-TOP**; and Mr. F. C. Browne gives **SURFER**, in his list of "gunners' names," at Plymouth Bay, Forest and Stream, November 9, 1876. In Maine, at Winter Harbor, **GOOGLE-NOSE**, originally Goggle-nose, I presume; at Ash Point (near Rockland), Bath, Portland, Pine Point, and Kennebunk, **PATCH-HEAD**; in Massachusetts at Fairhaven and New Bedford, to many upon Martha's Vineyard, and at Stonington, Conn., **PATCH-POLLED COOT**; at Bridgeport, Conn., to some of the gunners at least, **WHITE-SCOP** (referring to the white of the head, "scop" being old English for head or scalp). To some at Kennebunk, Me., **MUSCLE-BILL**; in Massachusetts at Salem, **PICTURED-BILL**; at Chatham, **PLASTER-BILL**. In Conn., at Stony Creek and Milford, **SNUFF-TAKER** (the drake's variegated beak reminding duckers of a careless snuff-taker's nose); at Stratford, **SPECKLED-BILL COOT** and **SPECTACLE COOT** (this latter name like Goggle-nose, the patches of black, one at either side of the bill, being likened to colored spectacles); Giraud writes (1844): "**SPECTACLE DUCK**, as it is by some called." At Bellport, L. I., **MOROCCO-JAW** and **WHITE-HEAD**. In New Jersey at Tuckerton, **BAY-COOT**; at Pleasantville (Atlantic Co.), **BLOS-SOM-BILL** and **BLOSSOM-HEAD.**

Audubon speaks of its being known to "the gunners of Long Island and New Jersey" as the **BLACK SEA-DUCK**, stating also that in Maine and Massachusetts it is "best known by the name of **BUTTER-BOAT-BILLED COOT.**" A shorter form of this latter title, viz., **BUTTERBOAT-BILL**, is given by De Kay (Zoölogy of New York, 1844), but I have never heard of these forms in actual use. De Kay also credits the species with **BOX COOT**; and we read in Water Birds of North America, of its being known "to some" in New England as **HOLLOW-BILLED COOT** (see No. 30).

The females and young males are, by many, regarded as a species distinct from the adult drakes; the two former being known on Buzzard's Bay, from New Bedford to Westport, by the name **PISHAUG**, and *very generally* along our coast as **GRAY COOT**, and less frequently **BROWN COOT**. (Species No. 30 is also popularly divided in like manner, while the females and young of No. 28 are, as a rule, correctly placed; the white wing-mark revealing the relationship.) This mistake is one very easily made, so different in appearance from the old cocks are these gray-brown birds; a majority also of those that come to us in the fall are young, therefore tamer, inclined to frequent the inlets, mouths of rivers, ponds, etc., and when shot are so much easier to pick, and on the table so much more tender and palatable.

See note preceding No. 28, for name Booby, etc.

A supposed "variety" of the species, called "Trowbridge's Surf Duck," "Long-billed Surf Duck," etc.. has been latterly eliminated; found to be, in other words, nothing more nor less than this bird.

No. 30.

Oidemia americana.

Adult male. Plumage black throughout, or practically so, the upper parts glossy and slightly iridescent, the lower parts having a more or less brownish cast. Bill (its upper division) without noticeable encroachment of feathers, but with hump as

No. 30. Adult Male.

shown in picture, and large patch of orange changing to yellow above; this patch extending from base to front of nostrils and including hump; remainder of bill uniformly blackish, as are legs and feet.

Female. Much smaller than adult male; plumage dusky grayish brown, paler or more or less mixed with dull white about throat, lower part of head, lower breast, and belly. Bill

No. 30. Female.

with no hump, and plain blackish (or an almost uniform mixture of gray and black). Legs and toes brownish gray, the webs black.

Young male. Closely resembling female.

Measurements about those of No. 29, but female of present species often falling short of lowest figures there mentioned.

Range, as given in A. O. U. Check List: "Coasts and larger lakes of northern North America; breeds in Labrador and the northern interior, south in winter to New Jersey, the Great Lakes, and California."

AMERICAN SCOTER: AMERICAN BLACK SCOTER: formerly believed identical with very similar European species (*Oidemia nigra*) and referred to simply as **SCOTER DUCK, BLACK SCOTER, BLACK DIVER,** etc. (without the prefix " American ").

In Maine, at Eastport, Millbridge, Frenchman's Bay, Ash

Point (near Rockland), Bath, Portland, and Pine Point, at Portsmouth, N. H., in Massachusetts at North Scituate, Barnstable,
Fairhaven, New Bedford, and Falmouth, and at Stony Creek,
Conn., BUTTER-BILL. In Maine at Machiasport, Jonesport, Mill-
bridge, and Kennebunk, and at Plymouth, Mass., YELLOW-BILL.
In Massachusetts at Pigeon Cove (Cape Ann), BUTTER-NOSE,
at North Plymouth, Fairhaven, and New Bedford, COPPER-NOSE
and COPPER-BILL, and at Edgartown, PUMPKIN-BLOSSOM COOT.
In Massachusetts at Salem and Cohasset, at Stonington, Conn.,
and on Long Island at Shinnecock Bay and Bellport, BLACK
COOT; the female (and young) being known at Salem as SMUTTY
COOT, at Chatham, same state, as FIZZY, and at Bellport and
Moriches, L. I., as BROAD-BILLED COOT. Of the species as a
whole, De Kay says (Zoölogy of New York, 1844): " Known on
this coast under the name of Broad-billed Coot, and farther east
by the name of Butter-bill." To some at Cohasset, BLACK BUT-
TER-BILL, and at Stony Creek, Conn., WHISTLING-COOT; at
Hudson's Bay, according to Fauna Boreali-Americana, 1831,
WHISTLING DUCK (No. 23 being the " whistler " of people gen-
erally).

I am credibly informed that in the vicinity of Rangely Lake,
Me., this bird is the SLEIGH-BELL DUCK; and according to
Water Birds of North America, it is called the HOLLOW-BILLED
COOT on "the Atlantic side of Long Island," this being "a desig-
nation applied in New England exclusively to the Surf Duck"
(No. 29)— I have myself never heard the name used for any
fowl.

The females and young (similar in appearance) are almost invariably regarded by duckers as a species distinct from the old
males, and though locally distinguished from the latter by names
previously mentioned, they are very generally classed under that
of GRAY COOT (see No. 29), and less commonly BROWN COOT.
The flesh of the young is highly esteemed by gunners, and, it
may be added, by almost every one who has ventured to try it.

See note preceding No. 28.

Erismatura rubida.

Body broad and flat; neck, wings, and legs short; feet large; bill almost as noticeably broad at end as Shoveller's, No. 14; tail of stiff pointed feathers, wedge-shaped, and often cocked up comically in the air.

The full-dressed drake very showy; sides of head below eyes white to throat; top of head, and the nape bright black; upper

No. 31. Adult Male ("full dress").

parts of body, with sides and neck, rich brownish red or mahogany color; wings and tail brownish black; lower plumage silver white waved with dusky gray. Feet bluish gray with dark webs. Bill blue.

As usually found, however, the bill and feet are darker, and the plumage practically that of the female, viz.: upper parts blackish, intermingled with dull reddish brown; the lower and lighter part of head (see picture) grayish white with a dusky bar running back from bill. Lower parts of body similar to plumage

No. 81. Female.

first described, but very much duller in tone. Indeed, in this more common dress, the species has a cheap, soiled, and "shop-worn" appearance.

Length about sixteen inches; extent twenty-two to twenty-three inches.

Range: North America in general.

Of its breeding-habits I know personally very little. Dr. Coues says: " Breeding from northern border of United States northward." A. O. U. Check List says: " Breeding throughout most of its North American range." Professor Ridgway's Manual (of 1886) does not mention its breeding-grounds.

Though this duck is a gourmand, and greatly inclined to obesity, it is as quick a diver as any known species. When wounded it pluckily struggles to escape to the last gasp, bleeding all the time like a prize pig. I hear of its being sometimes undone by a too bountiful food supply. Gunners near the mouth of the Maumee River tell of finding these self-indulgent little creatures floundering helplessly fat on the water, and in certain seasons floating about in a dying condition, or dead, in considerable numbers.

RUDDY DUCK of Wilson, 1814. Though first introduced in that year to the ornithologists, and as "very rare," the species may have been familiar enough to the gunners under one or more of the following names. It is difficult to believe that a variety now so very common, and mentioned by Dr. Sharpless in Doughty's Cabinet, Vol. III., 1833, as abounding "in every nook and cove" of the Chesapeake, was really as rare as Wilson supposed, though it has, doubtless, increased in numbers since his time.

At Machiasport, Me., **BLUE-BILL** (see Nos. 17, 18, 19); at Bath, Me., and Newport, R. I., **BROAD-BILL** (again, see Nos. 17, 18, 19; also No. 14); at Fairhaven, Mass., **BROAD-BILL DIPPER**; at Stonington, Conn., **HARD-HEADED BROAD-BILL**; in New Jersey at Barnegat, Tuckerton, and Atlantic City, **SLEEPY BROAD-BILL**; at Kennebunk, Me., **HORSE-TURD DIPPER** (the birds being so termed, I am told, from their habit, when alarmed, of huddling together in a mass); at Provincetown, Mass., **DIPPER, DOPPER,** and **DAPPER** (see No. 24); at Eastville, Va., **MUD DIPPER**; at Portsmouth, N. H., **BUMBLE-BEE COOT**; in Mass., at Cohasset, **CREEK COOT**; to some at Cohasset, and commonly at North Scituate (same state), **HORSE-TURD COOT**; at Baltimore, Md., **COOT** simply (see our Coot of the ornithologists No. 32; also Nos. 28, 29, 30, 33, 37); to some in the vicinity of Plymouth, Mass., **SPOON-BILL** (see No. 14): in the neighborhood of Niagara Falls, **SPOON-BILLED BUTTER-BALL**; occasionally at Havre de Grace, Md., Norfolk, Va., Newberne, N. C., Savannah, Ga., and commonly in Golden City, Mo., Palatka and Sandford, Fla., **BUTTER-BALL,** the commonest name at Norfolk being **BUTTER-DUCK.**

The Buffle-head, No. 24, may have a prior claim to "Butterball" and "Butter-duck," but how would it do to leave the Ruddy in full possession of all the names having butter in them, and to call the former, which is less valuable for table use, the *Oleomargarine*-ball, etc., etc.?

We hear also at Norfolk **BUTTER-BOWL, BATTER-SCOOT,** and **BLATHER-SCOOT,** and in the Norfolk *Virginian* of December 12, 1884, the species is referred to as **BLATHERSKITE** and **BLADDER-**

SCOOT. At Cohasset, Mass., and Newberne, N. C., **SLEEPY-HEAD**; in New Jersey at Pleasantville (Atlantic Co.), **SLEEPY-DUCK**; at Pleasantville, Atlantic City, and Somers Point, **SLEEPY COOT**; at Crisfield, Md., **SLEEPY BROTHER.**

In the vicinity of Detroit, and at Vienna Marsh, north of Toledo, the book-name "Ruddy" has taken quite a hold even among the market-gunners (the example of city sportsmen of course). It is always a surprise to meet one of these *authorized* names in actual service, particularly one like this, descriptive of a state of plumage that the gunners are least familiar with. Others at Detroit, and the "punters" of St. Clair Flats, refer to the species still as **FOOL-DUCK**, **DEAF-DUCK**, and **SHOT-POUCH** (the latter—considering the bird's ability to carry away shot—being far from inappropriate). Commonly known at Chicago, and in the Putnam Co. portion of the Illinois River, and by some at Norfolk, Va., as **BULL-NECK** (see Nos. 15, 17); less commonly at Chicago, and more facetiously as **STUB-AND-TWIST.**

Since finishing the list of names heard by myself in more northern localities, Mr. Henry. P. Ives, of Salem, Mass., a gentleman who is well acquainted with this species, tells me of hearing it commonly called the **DAUB-DUCK** at Rangely Lake, Me.

In the vicinity of Plymouth, Mass., **GOOSE WIDGEON**; at West Barnstable, same state, **WIDGEON COOT**, or **WIDGEON** simply (see our Widgeon of the books, No. 8; also Nos. 9, 12, 13, 17). In Massachusetts at Falmouth and Martha's Vineyard, in Connecticut at Stonington, East Haddam, mouth of Connecticut River, Wilmington, N. C., and Savannah, Ga., **HARD-HEAD**; to some at Martha's Vineyard, **TOUGH-HEAD.** At Newport, R. I., Stratford, Conn., very generally on Long Island, and at Norfolk, Va., **BOOBY** (see note preceding No. 28); and sometimes on the south side of Long Island, **BOOBY COOT.**

"Looby" has also been recorded as a name for this species (Zoölogy of New York, 1844, and elsewhere). I am inclined to believe, however, that it originated in the index of Giraud's Birds of Long Island, and is a misprint for *Booby*. If a mistake, it was a happy one, the two terms being synonymous.

At Red Bank (Monmouth Co.), N. J., **SALT-WATER TEAL**,

and Giraud, 1844, speaks of its being known by this name to
gunners of Chesapeake Bay; in St. Augustine, Fla., **BROWN
DIVING TEAL.**

In the vicinity of Philadelphia, at Somers Point, N. J., to
some at Washington, D. C., and at Savannah, Ga., **STIFF-TAIL**; at
Tuckerton, N. J., **QUILL-TAIL COOT**; at St. Georges, Del. (Dela-
ware and Chesapeake Canal), and to some at Havre de Grace,
PIN-TAIL (the Pin-tail duck of books, &c., No. 13, being the
"Sprig-tail" in these localities); called also **BRISTLE-TAIL** at
St. Georges, and referred to in an article on "Chesapeake duck
shooting," by Dr. I. T. Sharpless, Cab. Nat. Hist., Vol. I., 1830
("Doughty's Cabinet"), as **HEAVY-TAILED DUCK.** In index to
Giraud's Birds of Long Island, **STICK-TAIL**; in Turnbull's Birds
of East Pennsylvania and New Jersey, **SPINE-TAIL**; at St. Au-
gustine, Fla., **DIP-TAIL DIVER**; in De Kay's Zoölogy of New
York, **DUN-DIVER**; in Samuels's O. and O. of New England,
RUDDY DIVER; and Nuttall (1834), speaks of its being "common
in the market of Boston," and "generally known" as **DUN-BIRD.**

At Manasquan, N. J., **HICKORY-HEAD**; at Havre de Grace,
Md., **GREASER**, this being the commonest name here for the
species; and William Wagner, a well known Washington gunner, tells of hearing it called **WATER-PARTRIDGE,** and **STEEL-
HEAD**, on the Patuxent River, Md. (their Partridge being Bob-
white, No. 42); in the markets of Washington the Ruddy is
known as **ROOK.** Just think of it, a duck called a rook under
the very shadow of the Smithsonian.

At Newberne, N. C., **PADDY** and **NODDY.** Any one familiar
with the species will understand why such terms as "noddy,"
"sleepy-head," "fool-duck," "booby," etc., are applied; for
though these ducks are clever enough after having been wounded
or thoroughly aroused by the slaughter of their companions,
they are exceedingly stupid at other times. If they have not
been recently fired at, they exhibit very little fear at the approach of a boat, and even after having been awakened from
their dreams by the report of a gun, they will sometimes fly in
a dazed manner directly towards the shooter, and alight again
within easy shot. Two of the names referred to in this con-

nection are generally given to birds that have no place in a list of this kind—"booby" belonging in the books and elsewhere to the gannets—genus *Sula;* and "noddy" to a Southern species of tern—*Anous stolidus.*

Another name at Newberne for the Ruddy, and a very popular one, is **LIGHT-WOOD KNOT.** "Light-wood" is a Southern name for very resinous or fatty portions of pine, commonly obtained from trees that have been "scraped" for turpentine. The knot of this "light-wood" is proverbially hard, and the appellation is therefore like "**hard**-head," "tough-head," "stub-and-twist," etc., and refers to the difficulty sometimes experienced in quieting these creatures. To put shot into a Ruddy is one thing, to *kill* him quite another matter.

In the neighborhood of Morehead, N. C., **PADDY-WHACK;** occasionally at Wilmington, same state, **DINKEY** ("Hard-head" being the common name); and one Wilmington ducker told me of hearing the Ruddy called **DICKEY** by certain South Carolina gunners,—"Don't you know," said he, "how, when they start, they go *dickey-dickey-dickey,* patting the water with their wings and feet?"

At Charleston, S. C., **LEATHER-BACK;** on the Savannah River (above Savannah), **DUMPLING-DUCK;** and on the Ogeechee River, Ga., **HARD-TACK.**

In 1885, while devoting myself particularly to the study of this species, it seems to have been unusually common. The late C. S. Westcott ("Homo") wrote from Philadelphia to Forest and Stream of Oct. 29: "The number of Stiff-tails that have come this year is beyond anything for years. Twenty-five to thirty per boat are the average returns each day below Chester." The same fall I was told by John Kleinman, of Chicago, who is not only a "crack shot," but a close observer of the habits of birds as well, that he had rarely if ever before seen Bull-necks so numerous. Mr. J. S. Atwood, of Provincetown, Mass., wrote me, Oct. 11 (1885): "These Dippers are very numerous at the present time in this locality, and gunners will get from twenty to thirty in a day. We never see them in salt-water." And Mr. Atwood wrote again during the same month, "Some gunners

8   .

have shot as many as seventy-five in a day." In this year also, I heard the duckers of Stony Creek, Conn., talking about a duck that had lately come among them in considerable numbers. The little stranger proved to be our friend the Ruddy. It had been occasionally met with in previous years, but not often enough to create general interest. I was several times asked whether it was good to eat or not, what its real name was, etc.

No. 32.

Fulica americana.

Principally of a dark bluish slate color slightly tinged about the back with olive brown, the head and neck black or blackish. Feathers beneath tail white; wings narrowly edged with white, the secondaries (viz., feathers growing from second bone of wing) broadly tipped with the same; also somewhat whitish on lower

Mud-hen. No. 32. Adult. Coot

part of body. While the wings are closed (tightly closed) the white of the secondaries is not visible, and the white edging to the wings is not easily discerned. Bill of adult white, though with three dusky spots forming an incomplete band about it near tip. The bill is continued backward upon the forehead by a thick gristly chestnut colored skin. This continuation, or "frontal plate," is easily indented by the nail, in freshly killed specimens, or moved about upon the bony structure which it covers. Bill of young bird dusky flesh-color tinged greenish towards the tip; the "frontal plate" but partially developed. Eyes carmine red. Legs yellowish green, or slate color with greenish tinge, and dark or inky about the joints. Toes furnished with broadly scalloped membrane.

Length fourteen to sixteen inches; extent twenty-four to twenty-eight inches.

A good swimmer, looking very duck-like on the water.

Weight of freshly killed adult in hand at this writing, twenty-one ounces.

Range, North America, from Greenland and Alaska to Central America.

Though this species cannot be regarded as particularly interesting to sportsmen, yet it is too intimately associated with duck and rail shooting to be omitted from the list. Its flesh has certainly not a good reputation with the community at large, though champions may be found for almost any variety. I have, for instance, heard the market-gunners and hucksters at Norfolk, Va., very loud in their praises of this bird; some indeed declaring it superior to Canvas-back. But, though this latter duck is usually much overrated, it brings the marketman too good a price to be often tested at his table. At Wilmington, N. C., they say that the present species is peculiarly delicious after having fattened upon the rice fields, and we all know how very much the food of a bird has to do with its quality. We should be thankful that when from want of better sport we slaughter fowl of this kind, people may be found ready and willing to relieve our bags and consciences.

COOT: CINEREOUS COOT: COMMON COOT: the only bird in the United States recognized as Coot by ornithologists (see Nos. 28, 29, 30, 31, 33, 37); more correctly termed **AMERICAN COOT,** our bird differing slightly from the common coot of Europe.

I have not fallen in with this species, nor heard gunners name it along the coast from the St. Croix to the Penobscot. From Bangor to Cape Cod Bay, on the Niagara River, at Lake St. Clair, in the vicinity of Chicago, and on the Illinois River, it is the **MUD-HEN** (see No. 33); and Dr. David Crary, a veteran sportsman of Hartford, Conn., tells of hearing it so termed in Benton Co., Oregon. (For other "mud-hens," see Nos. 35, 36, 43.)

Rev. J. H. Langille, in Our Birds in their Haunts, 1884, describes its manner of rising from the water, "gradually with a spatting, splattering noise," etc., adding, "very properly do the Western hunters call this bird the **SPLATTERER.**"

Again, from Bangor to Cape Cod Bay, **MARSH-HEN** (this being perhaps equally popular with "Mud-hen"), and Mr. Everett Smith speaks of hearing it called the **BLUE MARSH-HEN** in Maine.* (For other "marsh-hens," see Nos. 33, 34, 35, 36.)

To some at Salem, Mass., and more commonly at Newport, R. I., **MEADOW-HEN** (see No. 35); in Massachusetts at Provincetown, Buzzard's Bay, and West Barnstable, **POND-HEN,** and at Falmouth, **WATER-HEN** (see No. 33), and Gosse (1847) speaks of this latter term as so used in Jamaica—the name Coot being given there to the Florida Gallinule.

At Havre de Grace, Md., **MOOR-HEN,** so termed by all (see No. 33).

In Connecticut at East Haddam, and mouth of Connecticut River, and at Moriches, L.I., **PULLDOO,** a corruption of the French *poule d'eau* (water-hen). Audubon (Ornith. Biog., III., 1835), speaks of **POULE D'EAU** being applied in Louisiana to both this bird and No. 35, and adds concerning the present species: " In all other parts of the Union it is known by the name of Mud-hen and Coot." "*All other parts of the Union*" was far too

---

* Birds of Maine, *Forest and Stream*, 1882–83.

broad a statement, it is in good keeping, however, with much that has been written about bird names.

Mr. C. W. Beckham writes, in his Notes on the Birds of Bayou Sara, La., Bull. Nutt. Ornith. Club, July, 1882: " Known here by the Creole name of **POULET DEAN.**" (*Dean*—French *doyen*, the eldest, chief, or oldest-looking *poulet*, compared with those smaller water-hens or *poulets*, the gallinules—and rails perhaps.)

To some at Buzzard's Bay, Mass., and commonly at East Haddam, Conn., **SEA-CROW;**[*] to some at Stratford, Conn., and at Baltimore, Md., **CROW-BILL;** in New Jersey at Manasquan, Barnegat, and Tuckerton, Washington, D. C., Alexandria, Va., and Crisfield, Md., **CROW-DUCK.** Giraud (1844) speaks of its being known " in some sections" of Long Island, and at Egg Harbor, N. J., as **WHITE-BILL** and **HEN-BILL.** To a majority of the gunners at Stratford, Conn., it is the **PELICK.**

Known very generally in Virginia, and southward to Florida, and less commonly in latter state at Jacksonville, St. Augustine, and Enterprise, as **BLUE-PETER** (quite familiar to the older Floridians by this name); popularly known at Jacksonville, St. Augustine, Enterprise, and Sanford by book-name, " Coot "—No. 33, however, sharing this name more or less indiscriminately with the present species.

March, in his Notes on the Birds of Jamaica (1863–64), calls it **IVORY-BILLED COOT,** and I have a memorandum crediting it also with the name **MUD-COOT,** the locality, however, or source from which derived, having been carelessly omitted.

The species is also credited with the name Flusterer. In Wilson's Ornithology (where our bird is described as identical with European Coot, *F. atra*) the following note appears: " In Carolina, they are called *flusterers*, from the noise they make in

---

[*] A name given by many people along the coast from Cape May to Cape Charles, to Black Skimmer, *Rynchops nigra* (not included in this book); this is a long-winged gull-like bird with lower parts white, and legs red; beak black and red, and peculiarly compressed—" razor-billed ;" the upper mandible (upper division of bill) grooved to receive blade-like edge of much larger lower mandible.

flying over the surface of the water.—A Voyage to Carolina, by John Lawson, p. 149." Audubon writes: "The appellation of ·flusterers' given to it by Mr. Lawson in his History of South Carolina, never came to my ear during my visits to that state." And Nuttall speaks of the American Coot "fluttering along the surface with both the wings and feet pattering over it ;" adding, "for which reason, according to Lawson in his History of Carolina, they had in that country received the name of Flusterers." Now this is just what Lawson says in work referred to (1709): "Black Flusterers; some call these Old Wives; they are as black as ink, the cocks have white faces, they always remain in the midst of rivers, and feed upon drift grass, carnels or sea-nettles; they are the fattest fowl I ever saw, and sometimes so heavy with flesh that they cannot rise out of the water; they make an odd sort of noise when they fly. What meat they are, I could never learn. Some call these the great bald Coot." Lawson nowhere mentions the term "flusterers" alone, and advances no reason for the naming, and his acquaintance with the bird that always remained in the midst of rivers, and of whose meat he could never learn, was certainly quite limited. It is not improbable that he got names and species somewhat mixed, and I am inclined to believe that in the following extract from his book he alludes rather more to Coot than to Florida Gallinule (No. 33): "Blue-Peters—the same as you call Water-hens in England, are here very numerous, and not regarded for eating." The name Blue-Peter was probably then, as now, generally applied in the Carolinas to this more common and more *blue* water-hen, No. 32.

No. 33.

Gallinula galeata.

Adult. Bluish slate color with (in full plumage) sooty black head and neck; the head sometimes more brownish than black. In general appearance considerably like No. 32, though smaller.

No. 33. Adult.

Slate color of lower belly mixed with white; purer white beneath tail and on edges of wings, as in No. 32, but decidedly browner than the latter bird on back, tail, and portions of the wings, and with white stripes on certain long loose feathers of the sides, and no white on "secondaries." Brown of back, etc., of a chocolate tint, with occasional tinges of olive. Bill at its end pea-green, the remainder, including the leathery continuation covering forehead, red or nearly so. This continuation much more extensive than in No. 32, and terminating squarely (not in a point as in the latter bird). Legs pea-pod green, with dusky joints; this leg-color changed, however, next to feathering of thighs, to bright yellow and orange red; the feet without noticeable membrane, "clean toed."

Young. Showing until long after attaining full size but slightly developed "frontal shield" (as this leathery continuation of bill over the forehead is sometimes called). Bill with no red anywhere about it; its end green, but less bright than in adult; remainder of bill, including encroachment upon forehead, dark greenish brown. Head and neck with no true black. Plumage of lower parts considerably mixed with white. Legs, immediately beneath feathering of thigh, light orange green with no red.

Length thirteen to fourteen inches, or a little more; extent twenty to twenty-two inches. Weight fourteen ounces.

Found here and there from the British Provinces southward to southern parts of South America.

I did not originally intend to include this species, and have therefore omitted to gather its common names as completely as I would otherwise have done. Though its range is wide, it is numerous in comparatively few localities. It is found very closely associated with the American Coot, No. 32, but is seen much less often on the wing, or upon the open water. Its habit of keeping a greater part of the time out of sight in the tall marsh-grass, and its resemblance at a distance to No. 32, have kept its name from many "local lists." It is a better bird for the table than the Coot—"a heap-sight sweeter meat," as my colored boatman expressed it.

FLORIDA GALLINULE: AMERICAN GALLINULE: COMMON GALLINULE: SCARLET-FRONTED GALLINULE: WATER-HEN. (see No. 32).

In a list of the birds of Oneida Co. and its vicinity, New York
—Ralph and Bagg, 1886—this species is mentioned as " very com-
mon on the marshes of Seneca River," and locally known as
WATER-CHICKEN.

In Connecticut, at East Haddam, and Essex, at Havre de
Grace, Md., and to many at Enterprise, Fla., **KING-RAIL** (see
No. 34); and we read in Forest and Stream, October 2, 1879, of
its " Natural History Editor" killing the species on the meadows
of the Housatonic, the writer adding—" called in that state
(Conn.) the King-rail." The species is more common in Connecticut than many suppose. One of the young birds used in
my description was killed while rail-shooting at East Haddam;
a number having been shot there during the same month, September, 1886, and five of them certainly during one tide.

At Washington, D. C., **KING-ORTOLAN**, and less commonly,
MARSH-PULLET; at Alexandria, Va., **KING-SORA.** The name
King-ortolan is given by Coues and Prentiss (Birds of District
Columbia, 1861–62) as an alias of *Rallus elegans*, No. 34 ; but No.
33 is certainly *the* king-rail of the District now (1887), and it
may be added, is much more like an enlarged form of " ortolan "
No. 37.

In the vicinity of Lake St. Clair, about Chicago, and to some
at Enterprise, Fla., **MUD-HEN,** and a friend writes from latter
state as follows (being provided with a stuffed specimen): " At
Indian River I showed it to six men in one day, and each said
at once, ' that's a mud-hen.' On being asked if it were not a
' blue-peter,' they said no, that's a different bird. One man said,
' *There ain't no other name for that bird but mud-hen.*' "

In the western localities mentioned this species is numerous
enough to be thoroughly well known, and is commonly recognized as a much superior bird to No. 32. Yet many gunners
loosely use the term " mud-hen " for each. Others, however,
in these localities and on the Illinois River (in Putnam Co., at
least) who are more particular in such matters, distinguish the
Gallinule as **RICE-HEN,** and again at Detroit and other points
near Lake St. Clair, as **RED-BILLED MUD-HEN, MOOR-HEN** (see
No. 32), and **MARSH-HEN.** (For other " marsh-hens," see Nos.

32, 34, 35, 36.) At St. Augustine, Fla., many class the Gallinule indiscriminately with No. 32, as **COOT**, and Gosse (1847), speaks of No. 33 being known as Coot in Jamaica, where No. 32 is the "Water-hen." (For other "coots," see Nos. 28, 29, 30, 31, 37.)

Some distinguish the Gallinule at St. Augustine as **SUMMER-COOT**, and at Enterprise as **MUD-PULLET**, and again in latter locality, and at Sanford same state, as the Florida Gallinule or Gallinule simply, probably through the influence of sports-men from other parts. A majority, however, on this southern shore of Lake Monroe, term it Mud-hen. A bright-looking young fellow at Sanford, Fla., to whom I showed a freshly killed specimen, told me that he had always known it as the **BLACK GALLINULE**—the word "black" distinguishing it from the Purple Gallinule, *Ionornis martinica* (a smaller, less common, green and purplish-blue bird, not included in this book). I showed the same specimen to a Sanford negro who said, "Why, dat a coot," adding, after I had pointed out the difference, "Yes, but day both coots." Another darky broke into the conversation with "Naw, dat no coot, coot got a white bill, dat a marsh-hen." The name Marsh-hen, nevertheless, is usually applied at Sanford to No. 34.

Health-seekers from numerous regions have greatly demoralized the vernacular of cracker, negro, etc., in Florida, and nowhere can there be found a nomenclature more completely muddled. A single species will have perhaps a half-dozen aliases in a single neighborhood, each alias being familiar alone to the man from whom you hear it. I encountered great difficulties while collecting bird names along the more northern coast of Maine, but my labor up there was pure play compared with experiences in the far South. Many names heard at these extremities (the latter in particular) I have omitted, regarding them as simply, or little more than, individual oddities. Nevertheless, enough has been winnowed from the chaff to serve as a partial key to enigmas. I should perhaps add that the negroes of Florida are far more familiar as a rule with birds and beasts than the whites are, though this is saying but very little.

Rallus elegans.

Above blackish brown and yellowish tan, the feathers being edged with latter tint, and broadly striped along their centres with the former; top of head nearly plain dark brown, a whitish line from bill over eye; side of head varying from tan color to gray; front of wing deep reddish tan color, nearly plain;

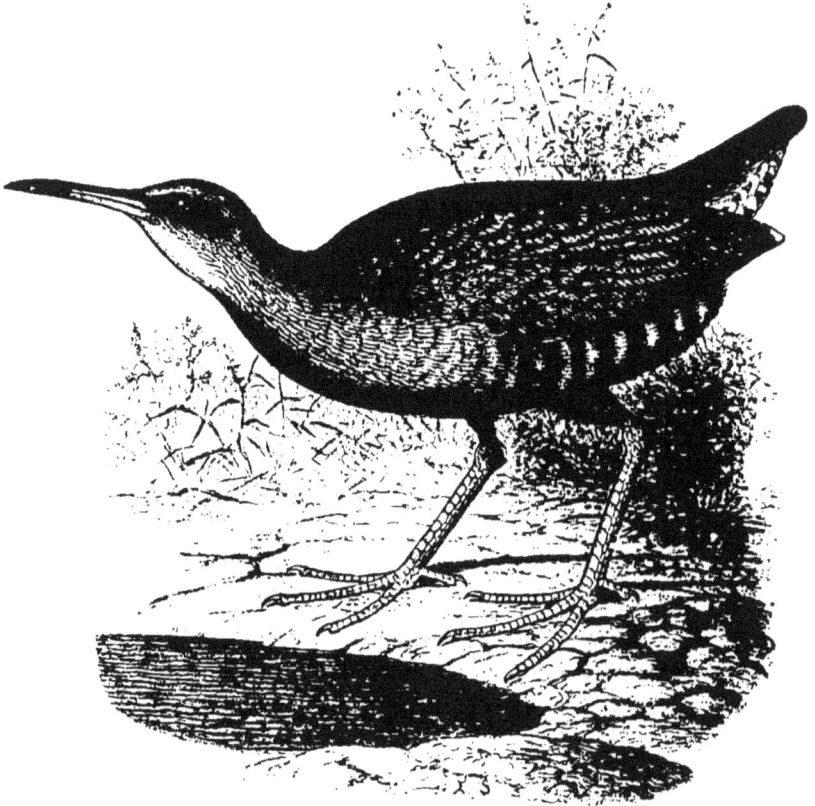

No. 34.

throat white; front of neck and breast nearly plain tan, or reddish cinnamon, this changing to a whitish mixture on lower surface of body; sides, flanks, and about thighs barred broadly with dark brown or black, and narrowly with white; the feathers immediately beneath the tail white, with touches of black and blackish brown. Bill blackish brown on top and at end, remainder brownish yellow. Legs yellowish brown with grayish olive tinge.

Length seventeen and a half to nineteen inches; extent twenty-three and a half to twenty-five inches; bill two and one eighth to two and a half inches.

Range, as given in A. O. U. Check List, 1886: "Fresh-water marshes of the Eastern Province of the United States, from the Middle States, Northern Illinois, Wisconsin, and Kansas southward. Casually north to Massachusetts, Maine, and Ontario."

KING RAIL: a name given also to No. 33, the present species, however, being the King Rail of "the books." Audubon speaks of killing "one female [of No. 34] in New Jersey, a few miles from Camden, in July, 1832," and "on inquiring of numerous hunters," was told "that they now and then obtained a few of these birds, which they considered as very rare, and knew only by the name of 'King Rails.'" (See No. 33 for name King Ortolan.)

Also termed, in print at least, **GREAT RED-BREASTED RAIL, FRESH-WATER MARSH-HEN** (see No. 36), **FRESH-MARSH HEN,** and **FRESH-WATER HEN.**

Very generally known throughout the South as **MARSH-HEN** simply, sharing this name, however, indiscriminately with the more common *Salt*-water Marsh-hen, No. 35, a similar bird, confused with the present species by many gunners, as it was, indeed, by Wilson,* the "father of American ornithology."

---

* Audubon, who exposed this confusion, wrote as follows: "No doubt exists in my mind that Wilson considered this beautiful bird as merely the adult of *Rallus crepitans* [No. 35], the manners of which he described, as studied at Great Egg Harbor, in New Jersey, while he gave in his works the figure and coloring of the present species. My friend Thomas Nuttall has done the same, without, I apprehend, having seen the two birds together."

No. 35.

Rallus longirostris crepitans.

Similar in general appearance to No. 34, but lower parts much less reddish—dull buff instead; upper parts more uniform in color, and more grayish or leaden in tone; the dark bars about thighs, flanks, etc., less dark; belly sometimes broadly white.

No. 35.

Bill practically as in No. 34. Legs gray, with yellowish or orange tinge about upper joint.

Length fourteen to sixteen inches; extent nineteen to twenty-one inches.

Frequenting salt marshes of Atlantic coast in large numbers from Long Island Sound southward, and occasionally found as far north as Massachusetts.

CLAPPER RAIL: MEADOW CLAPPER: SALT-WATER MEADOW-HEN: SALT-WATER MARSH-HEN: SALT-MARSH HEN: BIG RAIL.

In Connecticut at Stony Creek and Stratford, on Long Island at Bellport, Seaford, and Freeport, and to some on Cobb's Island, Va., **MEADOW-HEN.** Also at Freeport, and in New Jersey at Red Bank, Monmouth Co., Manasquan, Barnegat, Tuckerton, Pleasantville (Atlantic Co.), Atlantic City, Dennisville, and Cape May C. H., **MUD-HEN.** (For other "mud-hens," see Nos. 32, 33, 36, 43.) At Pleasantville above mentioned, at Pocomoke City, Md., on Cobb's Island, Va., and very generally to the southward, **MARSH-HEN.** (See *Fresh*-water Marsh-hen, No. 34, frequently confused with the present species, and termed also in many localities Marsh-hen simply. For other "marsh-hens," see Nos. 32, 33, 36.) Again at Pocomoke City, Md., and at Eastville, Va., **SEDGE-HEN;** very generally so called in these localities.

Gosse, in Birds of Jamaica, 1847, quotes an account of this rail written by his "friend, Mr. Hill," of Spanish-Town, in which it is stated that "the country people" call it the **MANGROVE-HEN,** and that "it greatly resembles the dappled gray variety of the common fowl," rambling about "with its callow brood, like a hen and chickens."

Rallus virginianus.

Upper plumage chiefly brownish black, the feathers edged with pale yellowish brown; side of head gray, blackening at base of bill; a light streak (grayish and pale buff) from near bill, backward over eye; top of head nearly plain blackish brown; wing in front reddish tan or mahogany color, with extreme fore edge of wing narrowly touched with pure white; throat whit-

No. 36.

ish; front of neck and front and sides of breast chiefly tan color; feathers of flanks black, narrowly barred at ends with white; feathers which cover from vent to end of tail variegated with black, white, and tan; thighs and along middle of breast gray, with whitish and light buff tints intermingled. Bill black above, and reddish flesh color below. Legs grayish brown.

Length eight and a half to nine and three quarter inches;

extent thirteen and a quarter to thirteen and three quarter inches; the bill, measured along its side, one and a quarter to one and a half inches.

Found here and there all along the coast, but met with oftener farther inland; a bird of the reedy swamp or marsh grass; widely distributed, but nowhere very numerous; though, perhaps, sometimes found fat, it has never been my luck to kill one that was not in a rather emaciated condition. I will add (as Mr. Sheppard has not shown the feet in his picture) that this bird's toes are free, like those of our other rails; that is to say, they are without webs or membranous attachments of any kind.

VIRGINIA RAIL: LITTLE RED-BREASTED RAIL: Wilson says, 1813 : " Known to some of the inhabitants along the sea-coast of New Jersey by the name of the **FRESH-WATER MUD-HEN**:" Nuttall, 1834, calls it **LESSER CLAPPER RAIL** and **SMALL MUD-HEN**: Giraud, in Birds of Long Island, 1844, speaks of its being "known to gunners and sportsmen" as **FRESH-WATER MARSH-HEN** (a name more commonly applied to No. 34). The late C. S. Westcott (" Homo ") describing " Rail Shooting on the Delaware "— Forest and Stream, Jan. 1, 1874—terms it **RED RAIL**, and states " that where fifty soras " (species No. 37) " are killed, but one or two red rails are boated."

In the vicinity of Salem, Mass., it is distinguished from the common rail, No. 37, as **LONG-BILLED RAIL**, but in most localities, in spite of longer bill, etc., it is loosely classed by gunners and marketmen with No. 37, under one of the latter's common names ; the difference between the species, however, being over-looked rather than unobserved.

9

No. 37.

Porzana carolina.

Adult. Above brown ("old gold" brown) with black on centres of feathers ; a narrow edging of white to feathers of fore-back and portions of wings. Bill yellowish green with much of its lower division bright yellow. Front of upper neck, throat, about bill, and streak running back on top of head, black ; sides of head and neck, and the breast, bluish slate color ; belly nearly

No. 37. Adult.

white; feathers immediately beneath tail, buff, or buff and white; sides of body, flanks, and lining of wings, barred white and brown, the brown replaced more or less on some individuals by black. Legs dull yellowish green.

Young. Without the black about head and throat, the slate color of breast, etc., being brownish in these regions, with dingy white throat. Upper parts generally of a lighter, more yellowish

No. 37.   Young.

brown. Lower surface of body light buff; the fore-breast similar to, or having the colors of, the upper parts. In other respects much as in adult.

Length eight and a half to nine inches; extent about fourteen inches.

Range, as given in A. O. U. Check List (1886): Temperate North America, but most common in the Eastern Province, breeding chiefly northward. South to West Indies and northern South America.

CAROLINA CRAKE: CAROLINA CRAKE GALLINULE: SOREE GALLINULE: LITTLE AMERICAN WATER-HEN: CAROLINA RAIL: AMERICAN RAIL: COMMON RAIL: SORA RAIL.

Known in New England very generally, and southward to New Jersey, as RAIL simply, this being occasionally heard here and there as far south as Florida. Giraud, in his Birds of Long Island, 1844, speaks of its being "known to the gunners by the name of ENGLISH RAIL," and De Kay, same date, Zoölogy of New York, mentions the latter name as one used in the South.

In New Jersey the gunners almost universally refer to the species as RAIL-BIRD, the word bird having been added, I suppose, to conform better with the name reed-bird,* these two species (the reed and rail birds) being commonly shot in one and the same reedy swamp, and together sent to market.

At Salem, Mass., it is distinguished from Long-billed Rail No. 36, as CHICKEN-BILLED RAIL or CHICKEN-BILL, and at East Haddam, Conn., it is the MEADOW CHICKEN (or Meadow Chick), the name Rail, though now in general use there, having been introduced by city sportsmen, who only a few years ago discovered that the East Haddam marshes were worth visiting.

At Washington, D. C., and Pocomoke City (Worcester Co.), Md., ORTOLAN (see foot-note concerning "reed-bird"). Generally known in Virginia and southward to southern part of North Carolina as SORA and SOREE. Catesby, in his Nat. Hist. Carolina, Florida, etc., 1731, gives this latter form (Soree), and Burnaby, describing travels "in the years 1759 and 1760," terms it SORUS, and speaks of meeting with the bird in Virginia during October, "at the tables of most of the planters, breakfast, dinner, and supper," and states in a note that "in several parts of Virginia the antient custom of eating meat at breakfast still continues."

In southern North Carolina, South Carolina, and Georgia, COOT, the bird being unrecognized by many by any other title. (For other "coots," see Nos. 28, 29, 30, 31, 32, 33.)

---

* Our Reed-bird—*Dolichonyx oryzivorus*—termed also Bobolink, Rice-bird, Skunk Blackbird, Ortolan, etc., is not included in this list. It is shot only for the "pot," having nothing more gamy about it than the English Sparrow has. I will add that I have nowhere found it called "Ortolan" but in print, and that the far-famed and delicious little Ortolan of Europe, from which the name is borrowed, is known to scientists as *Emberiza hortulana.*

The more common way of killing rail is familiar to many of us, viz., shooting them at high tide from the bow of a boat which is being poled through reeds and rushes by a man at the stern. In some localities, however, no powder is wasted. In the vicinity of Wilmington, N. C., for instance, the negroes, who do the greater part of the rail-killing, hunt them at night with pine torch and whip of brush-wood. The birds, interrupted at their supper of rice, wild oats, etc., are knocked down by this handful of brush, as they sit dazed by the light, or as they lazily attempt to wing their fat little bodies from harm's way. Neatly picked and tied in bunches, they bring the darkies from fifty to seventy-five cents per dozen.

9*

Tympanuchus americanus.

Male. Head slightly crested; on either side of neck a long
tuft of narrow feathers of uneven length, the longer ones black;

No. 38. Male (with neck-sacs inflated), and glimpse of Female.

beneath each tuft a bare spot of loose yellowish skin which the
bird has the power of inflating. Most of the upper plumage barred

transversely with dark brown varying to blackish and light tan color, the latter tint fading to white here and there. Under parts and sides regularly marked white and brown, in well-defined bars; throat buff. Legs covered to the toes with hairy feathers of drabbish tint, but more sparingly than in No. 39; the toes yellow brown.

Female. Similar to male, though with shorter, insignificant neck-tufts.

Length seventeen to eighteen inches; extent about twenty-eight inches.

This is the common pinnated grouse of Western prairies (and Eastern markets), regarded until very recently as identical with our once common Eastern variety which still exists upon Martha's Vineyard, Mass. Mr. Brewster, in the Auk of January, 1885, showed us that our Eastern bird (now known as *Tympanuchus cupido*) differs from the above-described Western variety, in being smaller, more reddish brown above, less white below, shorter legged, neck-tuft feathers "narrower and acutely instead of obtusely lance pointed;" the neck-tufts having also but four or five, instead of from seven to ten rigid feathers. Again, that our Eastern bird is "a *woodland* species, inhabiting scrubby tracts of oak and pine." This discovery—exceedingly interesting to scientists—is not important to gunners, the latter having practically nothing to do with the remaining handful of Eastern birds. "It is not unlikely," writes Mr. Brewster, "that the two forms intergraded over such intermediate ground as Western Pennsylvania and Eastern Ohio and Kentucky." How far to the east or west this intergradation extended it is, of course, impossible to tell. The names by which the two varieties have been known are as follows, no satisfactory separation being possible under the circumstances.*

PINNATED GROUSE: PRAIRIE HEN: PRAIRIE CHICKEN: (see No. 39): **HEATH HEN** (see No. 40), this being an early Eastern

---

* There is still another pinnated-grouse variety, found in the Southwest, and known in the books as *Tympanuchus pallidicinctus*, also as Texas Prairie Hen, Lesser Prairie Hen, and Pale Pinnated Grouse. •

name still applied in the vicinity of Martha's Vineyard to native birds, and to the Western grouse that have been introduced there. (I have one of the latter variety in my collection, shot at Falmouth, Mass., 1884). On Long Island—though no longer found there — pinnated grouse are still referred to as "heath-hens" by many of the older inhabitants. Dr. Mitchell in a letter to Wilson, dated 1810, concerning the Long Island birds, says: "Known there emphatically by the name of **GROUSE**" (see No. 39), adding, however, that "the more popular name for them is heath-hens."

Other old names are **BARREN HEN, HEATH COCK,** and **PINNATED HEATH COCK.** It was a common practice in early times to name our different grouse after "heath-game" of the old country. William Wood, speaking of our "birds and fowle" in New England's Prospect, 1634, says : "The flesh of the heathcocks is red, and the flesh of a partridge white;" and Daniel Denton, in A Brief Description of New York, 1670, tells of "heath-hens, quails, partridges," etc., as being found "in great store." Wilson relates a funny anecdote connected with the passage of a New York game law in 1791: "The bill was entitled, 'An Act for the preservation of Heath-hen and other game.' The honest chairman of the Assembly—no sportsman, I suppose — read the title, 'An Act for the preservation of *heathen,* and other game,' which seemed to astonish the Northern members, who could not see the propriety of preserving Indians, or any other heathen."

Pediocætes phasianellus campestris.

Without noticeable neck-tufts; tail-feathers graduated in length, the two central ones projecting as in picture; head slightly crested; hairy feathers covering legs down to between the toes; the toes gray. Upper plumage, in general, a closely

No. 39.

variegated mixture of buff, or grayish buff, tan color, and black, the buff tint sometimes paling to white here and there; the wing near body similarly colored, remainder of wing gray with spots and bars of white; throat buff. Lower parts, including wing-lining, chiefly white (snowy white), variegated about as in picture, with blackish brown U-shaped and V-shaped markings and touches of buff or light yellow-tan, the latter tint showing most noticeably along upper sides of body, where the markings are bolder and bar-like; the hairy feathers of the legs tinged with grayish buff.

Length seventeen to nineteen inches; extent twenty-six to twenty-nine inches.

This "chicken" is even more desirable, I think, than the pinnated kind (No. 38) for table use, and late in the season is a favorite in other respects, lying close to the dog, and jumping one or two at a time, instead of in a great "pack" a gun-shot away.

Range, as given in Mr. Ridgway's Manual, 1887: "Great Plains of United States, north to Manitoba (?), east to Wisconsin and northern Illinois, west to eastern Colorado, south to eastern New Mexico."

Until 1884 this grouse was regarded by every one as identical with *Pedioecetes phasianellus columbianus*, a variety whose range Mr. Ridgway gives as "Northwestern United States, south to northern California, Nevada, and Utah, east to western edge of Great Plains in Montana, north, chiefly west of Rocky Mountains (?) to Fort Yukon, Alaska."

Our more eastern form (*campestris*) differs from *columbianus*, according to Mr. Ridgway, in having the "ground color above more rusty or ochraceous." This difference, though interesting to naturalists, is something that gunners need not bother their heads with. The scientists themselves do not always agree about these very nice distinctions.

My list of names was prepared before hearing of the above distinction, but it will not be best to change it much now, the two forms having until so very recently been referred to as one and the same. Indeed, no satisfactory separation of these names is possible under the circumstances.

SHARP-TAILED GROUSE, or **SHARP-TAIL: PIN-TAILED GROUSE,** or **PIN-TAIL:** in Hallock's Sportsman's Gazetteer, 1878, **SPRIG-TAIL:** (No. 13, a duck, is also called Sharp-tail, Pin-tail, and Sprig-tail): PIN-TAIL CHICKEN: SPOTTED CHICKEN. In portions of our Northwest where the pinnated grouse (No. 38) are not found, this bird is the PRAIRIE CHICKEN; and Dr. Coues terms it PRAIRIE CHICKEN OF THE NORTHWEST. Popularly known also as WHITE-BELLY, and in some localities as WHITE GROUSE, the latter name immediately suggesting the ptarmigans (those grouse that turn white in winter), but to people familiar with our live sharp-tails, the word "white" seems rather appropriately applied, as the birds display so much of their white while flying. Mr. T. S. Van Dyke writes in a reminiscence of Minnesota shooting (Forest and Stream, Nov. 27, 1884): "Generally called the white grouse."

At Clarks (Merrick Co.), Nebraska, GROUSE (so termed, at least, in 1883), the pinnated species (No. 38) being the "prairie chicken." While shooting in the latter locality I was led to the conclusion that sharp-tails are rather more migratory than is generally believed. About November 10, 1883, during a cold storm, large numbers of these birds came upon us very suddenly. Not one had previously been seen, though several hunters, myself among them, had for a month or more been scouring the prairies almost daily.

Our United States sharp-tails have long been distinguished from the more northern and darker-colored variety *Pediocates phasianellus,*[*] by the names COLUMBIAN SHARP-TAILED GROUSE, COMMON SHARP-TAILED GROUSE, and SOUTHERN SHARP-TAILED GROUSE; and Mr. Henshaw, in an article headed "Prairie Chickens in Nevada," Forest and Stream, April 11, 1878, gives SOUTHERN SPIKE-TAILED GROUSE.

The "common" or English name designed expressly for *campestris* is PRAIRIE SHARP-TAILED GROUSE; this has certainly never been applied to any other sharp-tail; is, in other words, brand new.

---

[*] The range of this Northern Sharp-tailed Grouse (the "Long-tailed Grouse" of Edwards, 1750) is given by Mr. Ridgway as "Interior of British America, north to Fort Simpson, Fort Resolution, and Great Slave Lake, south to Moose Factory, Temiscamingue, Lake Winnipeg, and northern shore of Lake Superior."

Dendragapus canadensis.

Adult male. Head, neck, and back, barred with brownish black and slate-gray; wings light brown variegated with darker brown, gray, and occasional touches of white; tail-feathers very dark brown with light brown tips; throat black, skirted with speckled white; front of breast plain brownish black; sides of body variegated with grayish buff, dark brown, and white arrow-

No. 40. Adult Male.

like markings; lower surface blackish brown, variegated with white. A red membrane (or comb) over the eye. Legs feathered to the toes.

Female. Membrane over the eye less noticeable; tail barred, tan color and dark brown. Legs feathered as in male. General plumage a bright speckled mixture of light tan, dark brown, light gray, and white; the white not appearing upon the back proper, scarcely present about the head and neck, and showing most noticeably on the lower parts.

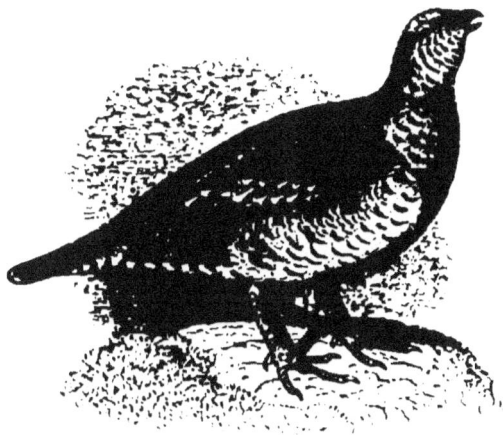

No. 40. Female.

Measurements about as follows: length fifteen to sixteen and a quarter inches; extent twenty-one to twenty-two inches.

This is not a well-known grouse to our gunners generally, as it is not met with very far south of our northern border. As a game bird it affords but little sport, comparatively, and its flesh, though sometimes good, at other times is disagreeably tinctured with the qualities of spruce buds or similar diet.

CANADA GROUSE: SPRUCE-GROUSE: SPRUCE PARTRIDGE: CEDAR PARTRIDGE: SWAMP PARTRIDGE: WOOD PARTRIDGE: WOOD GROUSE, (this term, like that of "timber-grouse," being sometimes broadly used to indicate grouse which inhabit woods, as opposed to those of the "open" or prairie): BLACK GROUSE: SPOTTED GROUSE.

At Eastport, Me., where the name Spruce-partridge is used to designate this bird, the Ruffed-grouse, No. 41, is called Birch-partridge, while at Jonesport, same state, the latter bird is Partridge simply, and the present species HEATH-HEN (see No. 38). This latter name suggests also that lengthy title applied in the middle of the last century (by Edwards) *i. e.*, **BLACK AND SPOTTED HEATHCOCK,** and it will be observed that the practice of christening our different grouse after "heath-game" of the old country was a very common one.

Bonasa umbellus.

The markings of this loosely feathered grouse are peculiarly complicated, and its prevailing tints variable; the difference between the grayest bird and the brownest, even in one and the same locality, being very great. Some may consider a description useless of a bird so well known as this. Yet I have myself shot with old and experienced gunners on Western prairies and Southern sea coast who had never seen a living or dead specimen of the species.

I will describe a pair of freshly killed adults now in hand.

Male. Upper parts reddish brown and gray, with touches

No. 41.

of dull black; on the back arrow-head or heart-shaped spots of light gray; the loose feathers of the neck (hackles) and those sweeping backward from shoulder regions boldly mottled reddish brown and black, and streakily marked with light gray; at either side of neck a glossy black "ruff" or "shoulder-knot," the feathers forming these being cut almost squarely at their broadened ends, and differing greatly from the long, narrow, neck-feathers of the pinnated grouse (No. 38); upper neck, back and sides of head, and crest, a speckled mixture of the bird's tints generally; tail light gray with faint ·tints of yellowish brown, broadly spreading like a fan, crossed by wavy black lines, and near its extremity by a broad black bar; much of upper plumage minutely flecked with black. Belly and breast mottled or brokenly barred, and feathering of flanks more broadly and decidedly barred with dusky brown and white; lower neck, sides of breast, and much of under parts tinged, or dulled, with light yellowish brown, a richer yellowish tint showing itself back of the vent; throat buff; front of neck crossed by narrow, dark brown, white, and bright tan markings. Toes, and naked part of leg just above them, gray; remainder of leg covered with hairy feathers of brownish white. Bill "horn color," dark above and light below.

Female. Throughout quite similar to male, but upper parts more brownish, chiefly, perhaps, about head and tail; the light arrow-head spots of the back washed with brown; the front of neck between the ruffs noticeably reddish brown or bright tan, without the narrow dark brown and white markings; neck-tufts very considerably smaller than in male, less cleanly black, and without noticeable gloss or lustre.

Measurements of species as frequently given: length eighteen inches; extent twenty-three or twenty-four inches. Dr. Coues, in his Key of 1884, gives length sixteen to eighteen inches; extent twenty-three inches.

The birds just described measure and weigh as follows: *Male:* length nineteen and three eighth inches; extent twenty-four and a quarter inches; tail from point of "pope's nose" to end of central feathers, six and five eighth inches; weight twenty-two ounces. *Female:* length seventeen and one eighth inches;

extent twenty-two and a quarter inches; tail (measured as before) a trifle less than four and seven eighth inches; weight
twenty-one and a half ounces.

RUFFED GROUSE: RUFFED HEATH-COCK (" Ruffed Heath-
cock, or Grous" of Edwards, 1758: see Nos. 38 and 40 for other
early applications of the name Heath-cock to our grouse):
BROWN RUFFED GROUSE:* **DRUMMING GROUSE: SHOULDER-
KNOT GROUSE** of Latham, who tells us (1783) of its being "called
by some the **DRUMMING-PARTRIDGE;**" and J. Sabine, in Appendix to Franklin's Journal, 1823, speaks of the name Shoulder-
knot Grouse as "well known to the British settlers in the northern parts."

" Frank Forester" says: "Properly called the Ruffed or **TIP-
PET GROUSE,**" and further remarks, " It is, therefore, equally
unsportsmanlike and unscientific to call the bird pheasant or
partridge; and it is, moreover, as needless as it is a stupid bar-
barism, since the bird has an excellent good name of its own,
by which it should invariably be styled, whether in writing or
in conversation, by every one claiming to share the spirit of the
gentle sportsman."—American edition of Hawker, 1846. "*Gentle*
sportsman" sounds funnily after such a tirade, and we could
smile broadly at the whole thing, had not this author's teachings
done so much to demoralize genuine young lovers of out-door
sport.

To some Canadians, **WHITE-FLESHER**; and this name is re-

---

* Dr. Coues so distinguishes it (Birds of the Northwest, 1874) from other
varieties of ruffed grouse, which are now recorded in A. O. U. Code and
Check List, 1886, as follows: Canadian Ruffed Grouse—*Bonasa umbellus togata*
—found in densely timbered portions of northern Maine and the British
Provinces, west to eastern Oregon and Washington Territory; Gray Ruffed
Grouse—*B. u. umbelloides*—Rocky Mountain region of United States and British America, north to Alaska; Oregon Ruffed Grouse—*B. u. sabini*—Coast
mountains of Oregon, Washington Territory, and British Columbia. As Dr.
Coues says, the reader "may ignore the varieties unless he desires to be very
precise. They are merely geographical races of the same bird, differing a
little in color according to certain climatic influences to which they are re-
spectively subjected."

corded as "Anglo-American" in Fauna Boreali-Americana, 1831. Again in the British Provinces, and at Calais and Eastport, Me., **BIRCH PARTRIDGE**; and from this to Pennsylvania, **PARTRIDGE** simply (see No. 42). In the latter state and throughout the bird's southern range (to Georgia and Arkansas), it is the **PHEASANT**, though in Virginia and the Carolinas we sometimes hear it referred to as the **MOUNTAIN PHEASANT**.

In Jefferson's Notes on Virginia, edition 1788, the names " Pheasant " and " Mountain Partridge " are given as belonging to one and the same species, *i. e.*, " Urogallus minor, or a ki. of Lagopus;" and Bartram, in Travels through North and South Carolina, etc., 1791, mentions (page 286) " Tetrao urogallus, or mountain cock or grous of Pennsylvania;" and again (page 290), " Tetrao lagopus, the mountain cock, or grous." These quotations indicate, in spite of the confusing Latin, an early application of the word "mountain" to our mountain-loving Ruffed Grouse. Bartram, while describing an evening in the northwestern part of South Carolina, in the edition of his Travels just cited, doubtless refers to the same species, when he speaks (page 331) of "the wary **MOOR FOWL** thundering in the distant echoing hills."

In a Natural History of North Carolina, 1737, John Brickell, M. D. (a quack who stole almost all of his material from Lawson), speaks of our "pheasants" differing from those in Ireland, and being "rather better and finer meat;" "their flesh," he adds, " is good in hectick fevers, the gall sharpens the sight, and the blood resists poison."

No. 42.

Colinus virginianus.

Male. Principally reddish brown, but with touches of black, gray, buff, and considerable white. Stripe along upper part of eye, white or nearly so; throat-patch white, ending below against

No. 42. Male, and glimpse of Female.

black collar-like shading; the white also speckling sides of neck, mixing with reddish brown and numerous waved and V-shaped lines of black upon the breast and along the sides of the body, and showing broadly in the vicinity of lower belly. Bill black. Legs gray.

Length about ten and a quarter inches; extent fifteen to sixteen inches.

Female. Resembling male in general appearance, but a little smaller; throat-patch and line over eye yellowish brown or buff (instead of white); marking beneath throat-patch much less dark; reddish brown of head lighter; under parts of body less broken by markings. Bill blackish, with lower part at base flesh-colored.

BOB-WHITE, so termed in imitation of its whistle, and the species is perhaps more widely recognized by this name than by any other, and though the name has been generally regarded as belonging rather to the pet-name or nickname order, it is now dignified by the endorsement of the American Ornithologists' Union, in its Code and Check List, 1886. There are several other imitations (regarded by none as names), like "no more wet," "more wet," "more wheat," "buck-wheat," etc.

Other titles copied from early and late authors are **VIR-GINIAN PARTRIDGE, MARYLAND PARTRIDGE, AMERICAN PARTRIDGE, COMMON AMERICAN PARTRIDGE, VIRGINIAN QUAIL, MARYLAND QUAIL, AMERICAN QUAIL, VIRGINIAN COLIN, AMERICAN COLIN:** the word "colin" being a Mexican name (for birds of the "quail-partridge" kind) brought to the notice of European naturalists by the work of Hernandez on the Natural History of Mexico, 1628, best known by edition of 1651.

Just how far north this bird is found in the West I cannot say, but in New England it is seldom seen as far north as Maine. From these northern limits to as far south as New Jersey it is the **QUAIL,** and in Southern States the **PARTRIDGE** (see No. 41). "Frank Forester" (Herbert) covered the ground as follows: "Where the ruffed grouse [No. 41] is called a part-ridge, the bird of which we are now speaking [No. 42] is called a quail . . . where the ruffed grouse is called the pheasant, our bird becomes the partridge."— American edition of Hawker, 1846. To this rule there are now numerous exceptions. In New Jersey, for instance, in many places where No. 42 is called

Quail, No. 41 is Pheasant, and, whether for better or for worse, the name Quail is growing more and more widely into fashion and favor. Southern pot-hunters, as well as sportsmen, instantly recognize No. 42 as the "quail" in numerous localities where, in former years, the name would as soon have been associated with a buzzard as a bob-white. In a communication from Memphis to Forest and Stream, October 1, 1885, concerning this bird's nomenclature, the writer, "Coahoma," says: "I never heard the term 'quail' applied to it until after the war, when a large influx of Northern sportsmen brought the name with them. Some Southern sportsmen, rather of the 'dude' order, have come to affect that name, but it is generally regarded as an innovation."

Captain John Smith certainly thought this bird looked more like a partridge than a quail, for he says, in his description of Virginia, 1612, "Patrridges there are little bigger than our Quailes" (the "quailes" to which he referred being European, of course).

A variety found in Florida is now distinguished in the books as " Florida Bob-white " (*Colinus virginianus floridanus*). It is a little smaller than No. 42, with general plumage somewhat darker, and black markings of under parts broader. It is unnecessary to mention in this work the other representatives of the genus, such as the " Cuban Bob-white," " Texan Bob-white," etc.

10*

No. 43.
Philohela minor.

Prevailing tint of plumage tan color, variegated above with black and gray, and nearly plain below; tail-feathers tipped on under side with white; bill and legs grayish flesh color, the former becoming black at end; eyes black, situated high, and far back. Sexes alike, but female the larger.

Length ten and a half to twelve inches; extent sixteen to eighteen inches; length of bill about two and three quarter inches; weight five and a half to nine ounces.

Western range of this favorite of Eastern sportsmen, according to A. O. U. Check List, 1886, "to Dakota, Kansas, etc."

No. 43.

Dr. Coues says : " This is *the* game bird, after all, say what you please of Snipe, Quail, or Grouse." Yes, Doctor, either in the field or on toast.

AMERICAN WOODCOCK: this (its correct name) distinguishing it from the European woodcock, *Scolopax rusticola.* It has been likewise termed **LITTLE WOODCOCK** and **LESSER WOODCOCK**, being considerably smaller than the Old World species.

In Bartram's Travels through North and South Carolina, etc., 1791, **GREAT RED WOODCOCK.** (See " little woodcock " applied to No. 44.)

Though known very generally as **WOODCOCK** in populous regions, we should bear in mind that this name is applied by backwoodsmen and other country-folk to the Pileated Woodpecker—*Ceophlœus pileatus*—wherever that big red-crested bird of the tall timber is found. Many funny stories are told of sportsmen being led far into the woods by promises of good " woodcock " shooting, only to find at the end of their journey that woodpeckers were the birds referred to.

It may be added that the popularization of the name " woodcock " for No. 43 is quite a modern accomplishment. Almost any old man that you may ask concerning the truth of this statement will tell you that he never associated the name with species now in hand during his boyhood. Many old people will tell you that as children they knew this bird by the name of **SNIPE.** Not merely as *a* snipe, be it understood, but as *the* snipe, and our woodcock is the " snipe " still, in rural districts far too numerous to mention, the species being commonly referred to collectively as " snipes " in these localities. See our true snipe, the snipe of most city people, sportsmen, and market-gunners, No. 44.

In an article entitled Woodcock Shooting, in Cabinet of Nat. Hist., Vol. I., 1830, the author speaks of the present species as called **BIG SNIPE, RED-BREASTED SNIPE** (see No. 45), and **MUD SNIPE;** and " big snipe " is the popular appellation among the crackers and negroes at Jacksonville, Fla., though many of them recognize the bird by its name " woodcock."

Dr. Barton, in his Fragments of Nat. Hist. of Penn., 1799, mentions the species as the **COMMON SNIPE**, as well as " woodcock ;" and Frank Forester gives **BIG-HEADED SNIPE** and **BLIND SNIPE** as two of the names by which "country folks" know it.* The latter name is also mentioned by De Kay (1844) as used " in some parts" of New York State, and Mr. George A. Boardman tells me of hearing the bird so termed in the vicinity of Calais, Me. For the sake of those who are wondering why this bird should have been called " blind," I will state that in spite of its large handsome eyes, its sight is noticeably dull in the full sunlight.

Hallock, in his Sportsman's Gazetteer (1879), credits the species with **WOOD SNIPE,†** and Dr. William Jarvis writes of hearing it termed **WHISTLING SNIPE** and **MUD HEN** some ten years ago at Cornish, N. H. (For name " mud-hen " as applied to other birds, see Nos. 32, 33, 35, 36.)

Audubon speaks of its being known in New Brunswick as **BOG-SUCKER.** Frank Forester, in Warwick Woodlands and elsewhere, frequently refers to it as the **TIMBER-DOODLE,** and in Lewis's American Sportsman it is credited with the names **MARSH PLOVER** (see No. 51) and **WOOD HEN** ; the author adding that the latter title " is not often applied."

"Homo" (the late C. S. Westcott) says, in an article on Autumn Woodcock Shooting, Forest and Stream, Jan. 22, 1874, "In the counties of Carbon and Lehigh, of Pennsylvania, capital grounds for autumn cock-shooting can be found in the neighborhood of Easton, Mauch Chunk and Lehighton, and I may state here that it is useless to inquire of the natives of these parts of the whereabouts of woodcock; very few know it by that name. I have heard it called **SHRUPS** and **BOG BIRD** by some." The term **COCK** used by "Homo," though heard in some lo-

---

* *Graham's Magazine*, December, 1843, " A Day in the Woods."

† Since quoting the above name from Mr. Hallock, a friend writes me that he pointed out a stuffed woodcock to a colored servant from Loudoun County, Va., and asked him if he knew what it was, and that the man immediately replied that it was what they called Wood Snipe in his part of the country.

calities, is oftener met with in print (the expression "cocking" meaning woodcock-shooting sometimes, as well as rooster-fighting). I only remember hearing the name "cock" popularly applied to the species in hand at Detroit; the gunners and marketmen there use it quite commonly.

At Pocomoke City (Worcester Co.), Md., and Eastville (Northampton Co.), Va., **NIGHT PARTRIDGE**; in the first-named locality, however, it is more commonly termed the **HOOKUM-PAKE**; the latter name being imitative of its notes, or those notes uttered immediately after its well-known spiral flight, the imitation being more intelligible if written as follows: hookum, —— hookum, —— p-a-k-e.

To the darkies about Matthews Court House, Va., **MOUN-TAIN PARTRIDGE**, and though we commonly associate woodcock with bogs and low-lying land, we must not forget the good shooting we have sometimes had higher up, nor the fact that many of these birds retire for a time to the hill-tops each year. In this connection the following from Mr. George B. Sennett's Birds of Western North Carolina is interesting (Auk, July, 1887). He writes: "I saw a pair of these birds on the summit of Roan in a clump of balsams; the overflow from numerous springs which had their sources at this spot formed an open adjoining marsh of several acres; altitude fully six thousand feet. One or two pairs have been known to breed here every year."

Dr. G. B. Grinnell, in Century Company's Sport with Rod and Gun, tells us that the woodcock is known to some in the seaboard counties of Virginia as Night Partridge (a name already recorded), and also as **PEWEE,** and in portions of North Carolina as the **NIGHT PECK.**

In an article on woodcock in Minot's Land and Game Birds of New England, 1877, the writer says: "Every sportsman is familiar with those very small, wiry, compactly feathered, weather-tanned birds who appear in October, and who are called, perhaps locally, 'Labrador Twisters.'" The birds referred to are probably those that when once disturbed and not immediately brought to bag, whirl away with surprising velocity upon a flight often too long to follow. Some old gunners believe that these "whistlers," or "little whistlers," as they are called in western Massachusetts and portions of Connecticut, are late birds from the far North, and that their appearance is always indicative of the end of the season's flight-shooting; while others claim that our fall shooting is as apt to close with large birds as with small ones, and that these very quick little fellows are old male "ground-keepers" (native stock). I incline myself to the latter theory, and will add that as these home birds are not fatigued by a journey from the North, they are naturally in good condition for rapid and protracted flight, and for a successful one, knowing well the ins and outs of home swamps and hillsides.

In many localities remote from bird-dogs and city influences the woodcock, though present perhaps in goodly number, has no name at all, is never pursued, and when accidentally flushed is regarded with little interest, or as a quagmire creature unfit for the food of man.

Gallinago delicata.

Upper parts chiefly dark brown and yellowish tan, the tan color of tail more reddish; the markings forming lengthwise stripes along the back (while wings are closed), and striping the upper part of head; chin and cheeks whitish; neck near head brown and pale tan, the colors of the upper parts shading here completely around; breast, sides of body, thighs, and lining of

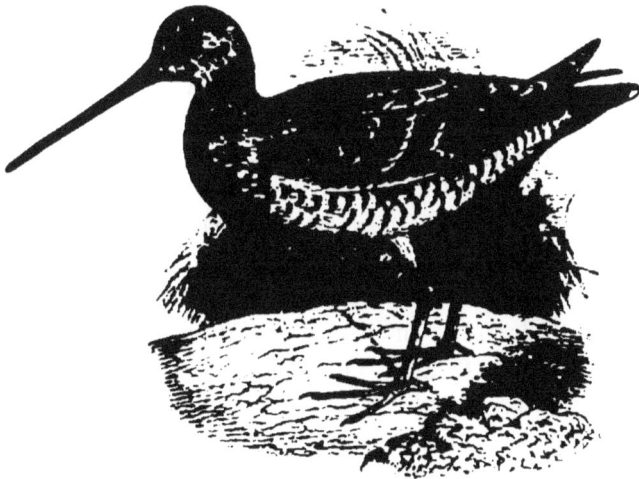

No. 44.

wings, white and gray, principally in fine bars; the belly white with a few touches of gray. Bill averaging about two and a half inches, and slightly flattened and spread near the tip, this flattened portion, in dried specimens, looking and feeling very much like a fine file; color of bill greenish gray near base, and blackish at the end. Legs light greenish gray.

Length ten and a half inches; extent about seventeen to eighteen inches.

A freshly killed bird now in hand weighs a little over four ounces, and its bill measures along the top a trifle less than two and three quarter inches.

WILSON'S SNIPE: AMERICAN SNIPE: COMMON SNIPE: SNIPE (see No. 43): mentioned in Bartram's Travels through North and South Carolina, etc., 1791, and in Barton's Fragments of the Natural History of Pennsylvania, 1799, as **MEADOW-SNIPE** (see No. 51); the latter author calling it also **LITTLE WOODCOCK.***

This favorite of our sportsmen and epicures is found throughout the United States. Its name **ENGLISH SNIPE** (the bird was regarded up to the time of Wilson as identical with European species) and that of **JACK SNIPE** (see Nos. 46, 51) I have not assigned to individual localities with any thoroughness, as both these names are so widely and popularly applied to it. A few scattering memoranda found among my notes are as follows: Known as *English Snipe* in Maine at Bangor, Bath, and Pine Point, in Massachusetts at Plymouth and Barnstable, in Connecticut at Stonington and Stony Creek, on Long Island at Moriches and Seaford, and in Florida at Enterprise and Sanford. Known as *Jack Snipe* at Portsmouth, N. H., in Massachusetts at Rowley and Salem, in Illinois at Chicago and in Putnam County, at Havre de Grace, Md., Washington, D. C., Alexandria, Va., and in Florida at Enterprise and Sanford (the two titles being about equally popular in the last two localities).

In New Jersey at Pleasantville (Atlantic Co.), Atlantic City, and Somers Point, **BOG SNIPE,** and at Crisfield, Md., **MARSH SNIPE.**

---

* At least there is no doubt in my own mind that Wilson's Snipe was the one referred to. Bartram mentions " *Scolopax Americana rufa,* great red woodcock," following it immediately with " *S. minor arvensis,* the meadow snipe." Barton gives " *Scolopax minor* (G), little wood-cock (meadow-snipe)," and again " *Scolopax minor, Scolopax minor arvensis* of Bartram, *Pi-si-co-lis?* of the Delaware Indians." In Zeisberger's Delaware Indian Spelling Book, 1776, we read, " *Me me u*—a Woodcock," and " *Pi si co lis*—a Snipe."

Nuttall (1834) speaks of its being known in the vicinity of Cambridge, Mass., as the **ALEWIFE BIRD,** "from its arrival with the shoals of that fish;" and Dr. Grinnell writes, in Century Magazine of October, 1883: "Few of our birds are so poor in local names as this one, for it is almost everywhere known either as the 'English' or the 'jack' snipe. Along the New England coast, however, it has an appellation which is rather curious. As the bird arrives about the same time as the shad, and is found on the meadows along the rivers where the nets are hauled, the fishermen when drawing their seines at night often start it from its moist resting-place, and hear its sharp cry as it flies away through the darkness. They do not know the cause of the sound, and from the association they have dubbed its author the **SHAD SPIRIT.**" Another and similar name associating this bird with the coming of the fish, is found in the following quotation from Krider's Sporting Anecdotes: "We have long noticed that when the nights are cool, with high winds from the northwest, towards the latter end of March, very few birds are to be found on the marshes. The prevalence of southerly winds and a hazy sky, with drizzling rain, is much more favorable to their migration northward. The same remark holds good in reference to the appearance of shad in the Delaware. Indeed, snipe are called **SHAD-BIRDS** by many of the fishermen, and the abundance or scarcity of the one is considered highly indicative of that of the other."

Mr. Ridgway tells me that the species is very commonly called **GUTTER SNIPE** in the southeastern part of Illinois, and he so terms it in his catalogue of the birds of Illinois, 1874.

In Wood's New England's Prospect, 1634, "snites" are mentioned among other birds, but we can only guess at the species referred to. Halliwell gives this old Anglo-Saxon name as "still in use" in parts of England in 1847, and in Drayton's Owl, 1604, we read of "the witless woodcock, and his neighbour snite," and in Baret's Alveary, 1580, "a snipe or snite, a bird lesse than a woodcocke."

The "simpes" mentioned in Morton's New English Canaan, 1637, perhaps meant *snipe,* but I cannot speak more confidently, as some have done, in regard to the matter. Josselyn, in his Voyages to New England, 1674, speaks of "widgeons, simps, teal," etc. If "simpes" and "simps" are synonymous, one can easily believe them to be of the duck kind, from the manner in which the latter author includes them.

Macrorhamphus griseus.

Adult in summer. Upper parts of body plumage (as viewed with wings closed) blackish brown, light cinnamon brown, and yellowish white, the latter tint paling to purer white here and there; shaft of longest wing-feather, and the lower back, white, the latter becoming mottled towards the tail; the tail itself

No. 45.

barred with blackish brown, light tan, and white; top of head and streak along line of eye dark (blackish brown predominating); sides of head and neck, and the under parts of plumage, light cinnamon brown, or buff, mottled with black; this black appearing more in the form of bars along the flanks and beneath the tail, where the feathers are bordered and otherwise marked with white. Bill greenish black, lighter at the base, and thoroughly *snipe*-like (the end being flattened and having little inequalities much as in No. 44). Legs and feet dull yellowish green, the outer and middle toes connected at base by a small, though noticeable, membrane.

Winter plumage. Light gray, nearly plain about the head, neck, and fore part of body; wings and shoulder feathers variegated with grayish brown and edgings of yellowish white: lower back white; the sides of head whitish, with a dusky line from bill through the eye; region about flanks and back of the thighs white; bill, legs, and feet as in summer.

Measurements about as follows: length ten and a quarter to eleven inches; extent seventeen and a half to nineteen inches; bill two and a quarter to two and five eighths inches.

Range, as given in A. O. U. Check List: " Eastern North America, breeding far north."

RED-BREASTED SNIPE (commonly so termed in the books; see Western Red-breasted Snipe, No. 46, found East occasionally): **GRAY SNIPE: BROWN SNIPE: NEW YORK GODWIT** of Swainson and Richardson, 1831.

In Maine at Portland and Pine Point, at Portsmouth, N. H., in Massachusetts at Salem, Provincetown, and West Barnstable, in New Jersey at Barnegat and Tuckerton, and on Hog Island, Va., **BROWN-BACK**; and Giraud mentions this name as common at Egg Harbor, N. J., 1844 (see No. 51). In Massachusetts at Salem, Rowley, Ipswich, in the vicinity of Boston and at Chatham, and in Connecticut at Lyme and Saybrook, **ROBIN-SNIPE** (see No. 52). At North Plymouth, Mass., **DRIVER.** At Stratford, Conn., and Seaford (Hempstead), L. I., **DOWITCHER;** on Long Island at Shinnecock Bay, Moriches, and Bellport, and at Barnegat, N. J., **DOWITCH** (see No. 51). These names Dowitch and Dowitcher meant originally that this was the Dutch, or German, snipe (*Duitsch, Deutscher*), and were probably employed to distinguish No. 45 particularly from the " *English*" snipe, No. 44. Giraud says, in his Birds of Long Island, 1844: " Our gunners, as if fearful that nothing would be left to connect the past with the present generation, cling to the old provincial names for birds, recognizing this species by the singular and unmeaning name of 'Dowitcher.'" It is interesting to note in this connection that the name Dowitcher is the one lately adopted by the American Ornithologists' Union, in their Code of Nomenclature and Check List. The name has been also written Dowitchee and Doe-witch. Giraud mentions also the name QUAIL SNIPE as used "in some sections of the Island;" and Colonel J. H. Powell, of Newport, R. I., writes (1885) of hearing it called GERMAN SNIPE on Long Island "some twenty-five years ago."

In New Jersey at Manasquan, Atlantic City, Somers Point, Cape May C. H., and Cape May City, in Virginia at Eastville, and Cobb's Island, GRAY-BACK (see No. 52); more commonly termed, however, at Cape May City, SEA - PIGEON. (It is scarcely necessary to mention that the latter is a guillemot name, as guillemots are not liable to be confused with birds that interest gunners and sportsmen.)

11

No. 46.
Macrorhamphus scolopaceus.

Very similar to No. 45 (one picture answering well for both), but the cinnamon brown or reddish tint of the summer plumage, covering—and more uniformly—the entire lower parts. Winter dress as in No. 45.

Length eleven to twelve inches; extent eighteen to twenty inches; bill two and a quarter to three inches.

Range, as given in A. O. U. Check List: "Mississippi Valley and Western Province of North America, from Mexico to Alaska. Less common but of regular occurrence along the Atlantic coast of the United States."

WESTERN RED-BREASTED SNIPE: GREATER GRAY-BACK: GREATER LONG-BEAK: LONG-BILLED SNIPE: RED-BELLIED SNIPE: LONG-BILLED DOWITCHER: WESTERN DOWITCHER.

Not being popularly recognized as distinct from the common Eastern variety No. 45, it naturally receives the latter's common names.

Mr. N. T. Lawrence says: "The gunners in the vicinity of Rockaway, L. I., make a distinction between the two birds, calling *M. scolopaceus* the **WHITE-TAIL DOWITCHER.**"—Bull. Nutt. Ornith. Club, July, 1880.

In the markets of Los Angeles, according to Dr. Cooper, **JACK SNIPE** (see Nos. 44, 51).

Symphemia semipalmata.

Adult in summer. Wings broadly marked with white and blackish brown, the white being on inner portions of the quill-feathers, and presenting (while wings are spread) a very conspicuous band across them; extreme lower part of back at base

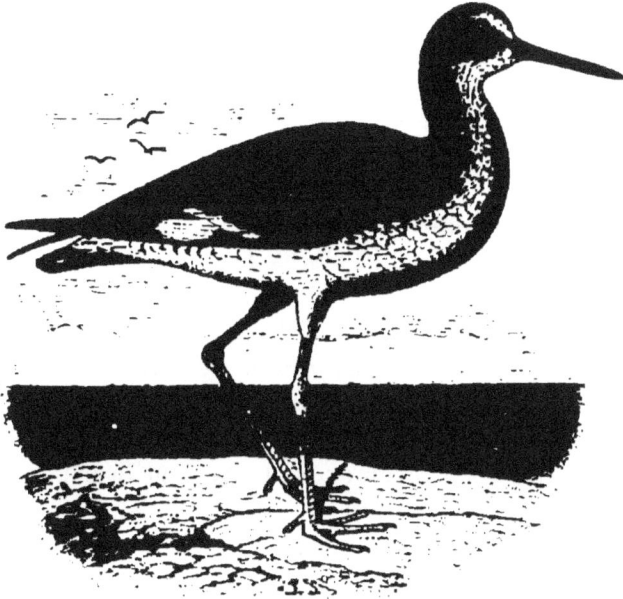

No. 47

of tail white; tail pale gray, more or less touched with dusky bar-like markings. Plumage in general light warm gray, mottled and shaded with brown and pale tan tints; the markings streaky about head and neck, spotty on the fore-breast, and in narrow, acutely waved bars upon the sides; markings much less numerous on the lower parts, leaving belly nearly white. Bill dark, paler towards base. Legs and feet dull gray, nails black.

Winter plumage. Upper parts chiefly pale warm gray nearly or quite free from markings; under parts wholly white, with the exception of some grayish shading on the lower neck; tail very pale, with white around its base. Wings, bill, and feet about as in summer.

Length fifteen and a half to sixteen and a half inches; extent twenty-seven to thirty inches. Bill two and one eighth to two and five eighths inches; toes with noticeable webs between them.

Range: "Temperate North America, south to the West Indies and Brazil" (A. O. U. Check List, 1886).

SEMIPALMATED TATTLER: SEMIPALMATED SNIPE.

On the coast of Maine this bird is too infrequently met with to be familiar to the gunners, and, indeed, the species does not occur abundantly anywhere in New England. We soon begin to find it common, however, as we move southward.

In Massachusetts at Rowley, Ipswich, Salem, Boston markets, North Scituate, Provincetown, North Plymouth, West Barnstable, Chatham, and New Bedford, **HUMILITY** * (see Nos. 48, 50); and Mr. Raymond L. Newcomb tells me of hearing it also called, at Salem, the **PIED-WINGED CURLEW.**

At Chatham, Mass. (to some of the gunners at least), at Newport, R. I., and southward along the entire Eastern coast, **WILLET,** though in Florida it is occasionally termed the **BILL-WILLIE.** In Lawson's Carolina, 1709, **WILL-WILLET;** elsewhere in print,

---

* On the Massachusetts coast this bird is sometimes confused with the Hudsonian Godwit, No. 61; I have heard, for example, the Godwit called "Humility," and a gentleman tells me of a Willet (No. 47) being sent him from Chatham, labelled with the Godwit's name "Goose-bird." As these occurrences, however, are simply mistakes, no further reference to them need be made. That gunners note a resemblance between the two species is instanced by the New Jersey name of "Carolina Willet" for the Godwit.

PILL-WILLET;* in Linsley's Birds of Connecticut, 1843, PILL-WILL-WILLET. (These last five names being imitative of the bird's shrill cries).

Audubon wrote of it as follows: "In the Middle States the Semipalmated Snipe is known to every fisherman-gunner by the name 'Willet,' and from the Carolinas southward by that of 'STONE CURLEW.'" Bryant, in his Birds of the Bahamas, 1859, speaks of its being known to the inhabitants as DUCK-SNIPE; and March says, in Birds of Jamaica, 1863–64, "Known here as the SPANISH PLOVER."

Mr. William Brewster writes, in the Auk of April, 1887, that Mr. J. M. Southwick has called his attention to the fact that Western specimens of the Willet differ from those of the Atlantic coast. The Western Willet, *Symphemia semipalmata inornata*, as Mr. Brewster terms it, differs from *S. semipalmata* in being a little larger, with "longer, slenderer bill;" and (in breeding dress) having the dark markings above "fewer, finer, and fainter," on a much paler ground, and those beneath duller, more confused, "and bordered by pinkish-salmon" which often "suffuses the entire under parts excepting the abdomen." In the winter dress the two forms "appear to be distinguishable only by size." Range (of Western variety): "Interior of North America between the Mississippi and the Rocky Mountains, wintering along the coasts of the Gulf and Southern Atlantic States (Florida, Georgia, South Carolina)."

---

* Mr. Dresser (cited by Baird, Brewer, and Ridgway) speaks of this name "Pill-willet" being applied by his boatman in Galveston Bay to the American Oyster-catcher—a bird seldom found north of New Jersey, and one which may be briefly described as follows, for the benefit of those who do not know it: Head and neck black; upper parts of body brown; under parts white; a white bar on the wing; and a red bill shaped for opening shell-fish. Again, in Bartram's Travels, 1791, we read of "the Will-willet or Oyster-catcher," and Audubon wrote of No. 47, "Its movements on wing greatly resemble those of the Oyster-catcher."

11*

No. 48.

Totanus melanoleucus.

Head and neck streakily marked with dusky gray and white, the throat nearly plain white; back and wings pale brown and blackish brown speckled with white and dull white, the colors of tail similar but forming bars; feathering just above tail white

No. 48.

barred with grayish brown. Under parts including lining of wings principally white, wavily and brokenly barred with grayish brown, the belly and neighborhood of tail nearly plain white. Bill blackish, two to two and a quarter inches long. Legs bright yellow.

Length thirteen to fourteen inches; extent twenty-three to twenty-five inches.

Found during its migrations throughout the country; flesh delicious in the fall, far better at this time than in spring.

GREATER YELLOW-LEGS: GREATER YELLOW-SHANKS: GREATER TELLTALE: TELLTALE: TELLTALE SNIPE: TELL-TALE GODWIT: TELLTALE TATTLER: VARIED TATTLER: LONG-LEGGED TATTLER: YELPER: YELLOW-SHANKS PLOVER.

Compare above names with those beginning list No. 49.

Wilson says (1813) of this bird and No. 49: "Well known to our duck-gunners along the sea-coast and marshes, by whom they are detested, and stigmatized with the names of the greater and lesser telltale, for their faithful vigilance in alarming the ducks with their loud and shrill whistle on the first glimpse of the gunner's approach." These birds are equally noisy and vigilant, however, when the ducks are absent, and they care very little about the welfare of other species than their own.

Called also **STONE SNIPE**, and in a communication to Forest and Stream of June 13, 1878, from Lebanon, Illinois, **STONE BIRD**.

In Maine at Eastport, Machiasport, Jonesport, Millbridge, and vicinity of Frenchman's Bay, Nos. 48 and 49 are both known as **YELLOW-LEG** or **YELLOW-LEGGED PLOVER**. At Machiasport, however, I am told by Captain James Robinson, the best-informed and one of the oldest gunners there, that many people in his locality apply the name **PLOVER** to this the larger species only, permitting the smaller one (No. 49) to monopolize that of Yellow-leg.

Audubon speaks of its being known in Maine as the **HUMILITY** (see Nos. 47, 50), adding that this is "an appellation that ill accords with its vociferous habits." In 1885 I made many inquiries in various parts of the state for this old name, but found only one man who remembered hearing it so applied, viz., the aged gunner Samuel Foote, residing near Bath, who referred to it as an appellation more or less common for the species during his youth, but very seldom, or never, heard now.

In Maine at Portland and Pine Point, in Massachusetts at

Rowley, Ipswich, Salem, North Scituate, Provincetown, North
Plymouth, West Barnstable, and to some at Newport, R. I., WIN-
TER (though originally an abbreviation of "Winter Yellow-leg,"
this is now a well-established title by itself).

At Portsmouth, N. H., in Massachusetts at North Scituate,
Fairhaven, New Bedford, and Falmouth, and at Shinnecock Bay,
L. I., WINTER YELLOW-LEG; and at Stonington, Conn., HORSE
YELLOW-LEG.

At Salem, Mass., the larger birds of the species have long
been distinguished from the others under the name of TURKEY-
BACK; some of the gunners there believing these biggest of
"big" yellow-legs a separate variety.

I have spoken of the name "Winter" as used "by some" at
Newport; the bird is commonly called there the BIG YELLOW-
LEG; and this, it may be added, is the name by which the bird
is best known throughout the country.

In Mr. Browne's list of "gunners' names," at Plymouth Bay
(Forest and Stream, November 9, 1876), LARGE CUCU; and a cor-
respondent ("Cohannet") in the same newspaper, December 2,
1886, speaks of its being known as CU-CU on the south shore of
Cape Cod.

In New Jersey at Dennisville, Cape May C. H., and Cape
May City, KILL-CU; this name being used by many of the gun-
ners for No. 49 as well. At Barnegat it is the Large or Big
Yellow-leg, though we hear the old name Telltale there occa-
sionally, and perhaps more often at Tuckerton and Atlantic City.
Generally termed in last two localities BIG YELLOW-LEGGED
PLOVER.

At Eastville (Northampton Co.), Va., YELLOW-SHINS, for
both this bird and No. 49.

Totanus flavipes.

Form, colors, and markings practically like No. 48 (the picture of the latter bird will do for both); its range also similar.

Length ten to eleven inches; extent about twenty inches; bill one and a half inches.

YELLOW-LEGS: COMMON YELLOW-LEGS: LESSER YELLOW-SHANKS: YELLOW-SHANKS SNIPE: YELLOW-SHANKS PLOVER: YELLOW-LEGGED GODWIT (so termed by J. Sabine, Appendix to Franklin's Journal, 1823): YELLOW-SHANKS TATTLER: TELL-TALE.

Compare above names with those beginning list No. 48.

In Maine at Eastport, Machiasport, Jonesport, Millbridge, and vicinity of Frenchman's Bay, YELLOW-LEG and YELLOW-LEGGED PLOVER (see No. 48); and in the vicinity of Henniker, N. H., PLOVER simply.

At Portland and Pine Point, Me., in Massachusetts at Rowley, Ipswich, Salem, North Scituate, Provincetown, North Plymouth, West Barnstable, and by some at Newport, R. I., SUMMER; the full name SUMMER YELLOW-LEG being commonly employed at Portsmouth, N. H., and in Massachusetts at Fairhaven, New Bedford, and Falmouth; often heard at North Scituate, and generally used at Shinnecock Bay, L. I.

At Plymouth Bay, according to Browne's list, 1876, SMALL CUCU; at Newport, LITTLE YELLOW-LEG—a name by which the bird is widely known throughout the country.

Though commonly known at Barnegat, N. J., as Little or Small Yellow-leg, the old name Telltale is heard occasionally, and we hear this latter title still more often perhaps at Tuckerton and Atlantic City, the bird being commonly distinguished, however, in these last two localities as **SMALL YELLOW-LEGGED PLOVER.**

Again in New Jersey at Dennisville, Cape May C. H., and Cape May City, **KILL-CU**; this name being loosely applied to Nos. 48 and 49, though used by some at Cape May City for No. 48 only.

At Eastville, Va., **YELLOW-SHINS** for both Nos. 48 and 49.

No. 50.

Bartramia longicauda.

Above, a mixture of blackish browns, the feathers edged with slightly reddish or rusty white; neck lighter, a yellowish brown with dusky streaks; sides of head light also; top of head dark brown; arrow-head markings about front of breast and lower

No. 50.

neck; throat and belly white with buff tints; inner surface of wings prettily barred gray and white; tail, a mixture of yellowish brown and white speckled and blotched with black. Legs light gray tinged with greenish yellow. Bill black above and at tip, the remainder bright yellow.

Measurements (derived from seven freshly killed birds):
length eleven and a quarter to twelve and a quarter inches;
extent twenty-one and three eighths to twenty-two and one
eighth inches; bill, measured along its top, about one and three
sixteenths inches.

BARTRAM'S SANDPIPER: BARTRAM'S TATTLER: UPLAND
SANDPIPER: BARTRAMIAN SANDPIPER: BARTRAMIAN TAT-
TLER: BARTRAM'S HIGHLAND SNIPE.* Wilson says, Vol. VII.,
1813: "This bird being, as far as I can discover, a new species,
undescribed by any former author, I have honored it with the
name of my very worthy friend [William Bartram], near whose
botanic gardens, on the banks of the river Schuylkill, I first
found it."

This is proverbially a difficult bird to approach; is found
throughout the country east of the Rocky Mountains, and is
a great favorite among sportsmen and epicures.

In Maine at Bangor, Rockland, Bath, Portland, and Pine
Point, at Portsmouth, N. H., in Massachusetts at Ipswich, Provincetown, and West Barnstable, UPLAND PLOVER.

Concerning its book-name, Bartram's Sandpiper, Mr. E. E. T.
Seton, of Manitoba, says in an article on popular names of birds
(Auk, July, 1885): "Ever since Wilson's time this name has been
continually thrust into the face of the public, only to be as continually rejected; Upland Plover it continues to be in the East,
and QUAILY on the Assiniboine."

At Bangor, Me., and in New Jersey at Barnegat, Tuckerton,
and Cape May C. H., FIELD PLOVER; and Dr. Wheaton writes
in a report of the birds of Ohio (Columbus, 1879): "Field Plover,
as it is commonly termed with us."

At Bath and Portland, Me., HIGHLAND PLOVER; at Portsmouth, N. H., and Salem, Mass., PASTURE PLOVER; at Provincetown, UPLANDER; in Maynard's Birds of Eastern Massachusetts, HILL BIRD; at New Bedford, Mass., Newport, R. I.,

* So termed by Dr. Woodhouse, Sitgreaves' Expedition, Zuni and Colorado
Rivers, 1853.

and Stonington, Conn., and to some at Shinnecock Bay, L. I.,
GRASS PLOVER; on Long Island at Shinnecock Bay and Moriches, HUMILITY (see Nos. 47, 48); at Bellport, GRAY PLOVER
(see No. 55); at Seaford (Hempstead), PLAIN PLOVER; and we
hear CORN-FIELD PLOVER among other names at Washington, D. C.

Dr. Coues, in Birds of the Northwest (1874), speaks of the
species as very numerous during its migrations "in most parts
of the West between the Mississippi and the Rocky Mountains,"
and "commonly known as the PRAIRIE PIGEON" (see No. 56);
and we find the following in Water Birds of North America
(Baird, Brewer, and Ridgway): "In Southern Wisconsin, Mr.
Kumlien informs me, in 1851 this bird, then very common there,
was known as the PRAIRIE PLOVER, and also as the PRAIRIE
SNIPE." Mr. Seton, previously cited, calls it also Prairie Plover
(as well as Quaily) in his Birds of Western Manitoba, Auk, April,
1886.

At New Orleans, La., it is the PAPABOTE; this is Audubon's
spelling of the name; it is also written "Papabot" and "Papa-
botte."

Concerning the name Frost Bird, credited to the species by
Herbert, see No. 56.

No. 51.

Tringa maculata.

Bill practically straight, though with a very slight downward curve, yellowish or dull yellowish olive at base, the remainder black; the two central feathers of the tail projecting beyond the others and more pointed; the top of head dark, a brownish mixture; throat white; sides of head, neck all around, and breast streakily grayish buff and dusky brown; the sides of the head

No. 51.

having an indistinct whitish and a dusky stripe. Upper plumage in general a mixture of light yellowish or reddish brown, with dark or blackish brown and a few touches of white. Lower parts (back of the breast-markings) white. Legs dull yellowish olive.

Measurements about as follows: length eight and a quarter

to nine and a quarter inches; extent sixteen to seventeen and a half inches; bill measured on top from feathers to tip one to one and three sixteenths of an inch.

Range wide, including during migrations all of North America; a good little bird for the table, and as a rule easily walked up to and shot where it stands, neck drawn in as though asleep. It will sometimes, however, mount into the air from concealment, and whirl away upon a snipe-like flight that is not easily stopped.

PECTORAL SANDPIPER (so termed in the books): **JACK SNIPE** (see No. 44, to which this name is more generally applied; also No. 46). As I have never happened to hear the latter title in use for this species, I must quote others concerning it. In Water Birds of North America (Baird, Brewer, and Ridgway) we find the following: " Mr. Boardman informs me that this species is quite common, both in the spring and in the fall, near Calais, Me., where it is seen in company with the Common Snipe, and where it feeds exclusively on the fresh-water marshes and in the uplands. It is distinguished from the Common Snipe by the name of the Jack Snipe." Mr. E. S. Bowler (Taxidermist), of Bangor, tells me that this name is so used in his locality. Giraud writes, 1844: " Mr. Baird has informed me that it occurs in Pennsylvania, in which section it has received the appellation of ' jack snipe;' " and in " Philadelphia Notes " to Forest and Stream, October 1, 1885, " Homo " (the late C. S. Westcott) says: " A few flocks of creakers, jack snipe they call them here, occupy the mud flats of the Delaware."

At Pine Point, Me., Portsmouth, N. H., in Massachusetts at Rowley, Ipswich, Salem, North Scituate, Provincetown, Plymouth, West Barnstable, Chatham, New Bedford, and Falmouth, **GRASS-BIRD**, and, infrequently, **GRASS SNIPE**. Known also to some at Rowley and Ipswich as **BROWN-BACK** (see No. 45); " X. Y. Z.," in Forest and Stream, November 18, 1886, speaks of its being " generally called **BROWNIE** " in the vicinity of Newburyport; and Mr. F. C. Browne, in his list of gunners' names at Plymouth Bay (Forest and Stream, November 9, 1876), gives **MARSH PLOVER** (see No. 43).

At Newport, R. I., on Long Island at Shinnecock Bay, Moriches, and Bellport, and at Barnegat, N. J., **KRIEKER.** I write this name as it is usually spelled. It was not applied, as popularly believed, because of the bird's creaking note, but because of its crouching or squatting habit—German *Kriecher*, a cringing person.

Known "to some of the residents" of Long Island (Giraud writes, 1844), as **MEADOW SNIPE** (see No. 44). At Essex, Conn., and mouth of Connecticut River, **DOWITCH** (a name belonging to the Red-breasted Snipe, No. 45, and interpreted under that head). In Connecticut at Milford, **SQUAT-SNIPE;** at Stratford, **SQUATTER.** At Seaford, L. I., **SHORT-NECK.** In New Jersey at Tuckerton, **FAT-BIRD;** at Pleasantville (Atlantic Co.), Atlantic City, and Cape May City, **HAY-BIRD.** Known also to some Atlantic City gunners as **TRIDDLER.** At Alexandria, Va., **COW-SNIPE.**

In Water Birds of North America the name "Crouching Shore-bird" is given as used at Trinidad. This (like Krieker or *Kriecher*, Squatter, etc.), is an appropriate appellation, but a translation, and a very free one it seems. Léotaud, in *Oiseaux de l'Ile de la Trinidad*, 1866, gives under the head of *T. maculata*, "*Vulg. Couchante;*" and Mr. Ridgway writes me that "this appears to be the only basis of Dr. Brewer's statement."

No. 52.

Tringa canutus.

Adult in spring. Upper parts a mixture of light buff, gray, black, dull brown, and grayish white; the feathers near the tail (upper tail coverts) barred with brownish black and white. Under parts uniformly reddish buff or orange brown, sometimes

No. 52.

fading to white on the lower part of belly; this reddish color much like that on the breast of our common garden robin.

Adult in autumn. Above almost uniformly gray. Below nearly white, having little or no robin color; front of neck, breast, and sides streakily freckled and otherwise flecked with dusky.

Young. Upper parts gray, with crescent-like dusky and

12

whitish borderings to the feathers. Lower parts white or nearly so, the breast sometimes tinged with buff. The neck and front of body streakily flecked, the sides faintly and irregularly barred, with dusky.

The different plumages of this species ("red," "ash-colored," and the variations between these) have caused some of its names to appear as very contradictory.

Length ten and a half inches; extent twenty inches or more; legs and bill nearly black, the latter one and three eighths to one and a half inches long.

A very good bird for the table, and well known to most of the world.

KNOT:* RED-BREASTED SANDPIPER: RED SANDPIPER: ASH-COLORED SANDPIPER: FRECKLED SANDPIPER: GRISLED SANDPIPER. The last two titles are given (among other names) by Latham, Syn., 1785. Giraud says, in his Birds of Long Island, 1844: " Late in September it moves southward; at this period the lower plumage is white, spotted on the neck, breast, and flanks with dusky ; the upper plumage ash gray ; in this dress it is the **WHITE ROBIN SNIPE** of our gunners." Wilson says: " The common name of this species on our sea-coast is the **GRAY-BACK**" (see No. 45), and we find the following in Audubon: " My friend, John Bachman, states that this species is quite abundant in South Carolina, in its autumn and spring migrations, but that he has never seen it there in full plumage. In that country it is called the **MAY-BIRD**, which, however, is a name also given to the Rice-bird. Along the coasts of our Middle District, it is usually known by the name Gray-back." In

---

* Canute, or Knut, king of Denmark and conqueror of England, was forced to retreat—we are told—before the incoming tide (in a manner to shame certain courtiers who claimed that the sea would obey him) even as this big sandpiper is driven by the waves, in common with smaller birds. It has been stated also, that this species was a great favorite with the old king—

"The *Knot* that called was Canutus' bird of old,
    Of that great king of Danes his name that still doth hold,
    His appetite to please that far and near was sought."—DRAYTON.

March's Birds of Jamaica (1863–64), it is the **WHITE-BELLIED SNIPE.**

At Pine Point, Me., and in Massachusetts at Boston markets, North Scituate, Provincetown, Plymouth, and West Barnstable, **RED-BREAST PLOVER**; in the above localities, and at Chatham, Mass., in Atlantic County, N. J., and at Eastville, Va., **RED-BREAST.** At Ipswich, Mass., **BUFF-BREAST, BLUE PLOVER,** and **SILVER-BACK.** At Newport, R. I., on Long Island at Shinne-cock Bay, Moriches, Bellport, and Seaford (Hempstead), in New Jersey at Barnegat, Tuckerton, Pleasantville (Atlantic Co.), and Cape May City, and at Eastville, Va., **ROBIN-SNIPE,** this being often shortened (particularly among the Long Island gunners) to **ROBIN.** Again at Moriches, L. I., and at Morehead, N. C., **BEACH-ROBIN**; at Manasquan, N. J., **ROBIN-BREAST**; and at Pleasant-ville above mentioned, **HORSE-FOOT SNIPE** (see No. 54).

No. 53.

Tringa alpina pacifica.

Bill with slight downward bend; the two central feathers of the tail a little longer than the others and more pointed; the bill and legs black.

Summer plumage. Above cinnamon brown, or light reddish brown, with elongated touches of black along the centres of the feathers; wings chiefly gray and dark brown, the feathers edged

No. 53. Summer Plumage.

and otherwise marked with white; sides of head, the neck, and breast, grayish white streakily marked with dusky; throat white; reddish tone of upper parts extending in greater or less degree up back of neck and over crown. Belly with large black blotch; remaining under parts white.

Winter plumage. Above plain warm gray; an indistinct

whitish streak along by the eye, from the upper part of the bill. Under parts white (or very nearly so) excepting lower part of neck and the fore breast, which are streakily grayish.

No. 53. Winter Plumage.

Length eight and a half inches; extent fifteen inches.

The range of this species includes our whole country; in the fall it is numerous along the sea-coast, often collecting in very large flocks. No apology is necessary for introducing it here; it has (notwithstanding its diminutive size) appeared many times in lists of gunners' birds; is plump and palatable in the autumn, and affords some sport even to adults, when bigger birds are absent. A record of its aliases may also prevent us from confusing it with other species.

DUNLIN: more correctly **AMERICAN DUNLIN** (to distinguish it from the European dunlin, *T. alpina*): **RED-BACKED SANDPIPER**: **AMERICAN RED-BACKED SANDPIPER**: **BLACK-BELLIED SAND-PIPER**: **BLACK-BREASTED SANDPIPER**: **PURRE** (Pennant, 1785): **OX-BIRD** (Nuttall, 1834).

Wilson (1813) speaks of its being known " along the shores of New Jersey" as the **RED-BACK**; and McIlwraith, in Birds of Ontario, 1886, mentions it as the "**BLACK-HEART PLOVER** of sportsmen;" and again as **BLACK-HEART** simply, and we find this latter form in E. E. T. Seton's Birds of Western Manitoba, Auk, April, 1886. In Forest and Stream, Nov. 18, 1886, " X.Y.Z."

12*

(Raymond L. Newcomb) tells of its being known in the vicinity of Gloucester, Mass., as **SIMPLETON**; and F. C. Browne gives **STIB** in his list of gunners' names at Plymouth Bay (Forest and Stream, Nov. 9, 1876).

At Pine Point, Me., Seaford, L. I., and Pleasantville (Atlantic Co.), N. J., **FALL SNIPE**; in Massachusetts at Chatham, **CROOKED-BILLED SNIPE**; at West Barnstable, **CALIFORNIA-PEEP**; at Newport, R. I., and in New Jersey at Tuckerton and Atlantic City, **WINTER SNIPE**;* at Stratford, Conn., and Shinnecock Bay, L. I., **LITTLE BLACK-BREAST**; at Seaford, L. I., in New Jersey at Tuckerton, Pleasantville, above mentioned, Atlantic City, Cape May C. H., Cape May City, and Cobb's Island, Va., **BLACK-BREAST** (see Nos. 55, 56); at Shinnecock Bay, **LEAD-BACK**; in New Jersey at Barnegat and Tuckerton, **BRANT-SNIPE**; and at Atlantic City, **BRANT-BIRD** (see Nos. 54, 60, 61). The gunners of the last-named localities claim that this little sandpiper is more closely associated than other birds with the Brant (No. 3); is more often found with the latter species on sandbars, sea-weed bunches, etc.

---

* This name is also applied to the Purple Sandpiper, *Tringa maritima*, a bird which comes down from the North in cold weather, is never seen by us before late autumn nor after the early spring, and whose appearance at this time may be briefly described as follows : very dark brownish slate color, showing purplish gloss in certain lights; belly white; length nine inches, or thereabouts; extent fifteen to sixteen inches; bill about one and a quarter inches, and nearly straight. Perhaps it would have been better to include this sandpiper more formally in my list, but it is practically an unknown bird to other of our gunners than those of New England (though occasionally found on the Great Lakes and elsewhere). The isolated bits of rocky coast which it inhabits are not inviting during wintry weather, and the bird is fallen in with generally by accident, for gunners are not on the lookout for shore-bird shooting at such times. Mr. George A. Boardman (cited by Baird, Brewer, and Ridgway), states that *T. maritima* is the Winter-snipe at Calais, Me.; and Mr. William Brewster tells me of its being so termed at Swamscott, Mass. I have heard it called Winter Rock-bird at Ash Point, Me. (the gunners there usually finding it at Green Island, ten miles southward); and it is the Rock-bird, Rock Plover, and Rock Snipe at Rowley and Salem, Mass.

The great difference between the winter and summer dress
has caused much confusion, and the " Black-heart," " Black-
breast," etc., is very generally regarded as a species quite distinct from the " Winter-snipe," " Lead-back," etc.

No. 54.

Arenaria interpres.

Adult male. Head, neck, and breast broadly pied black and white; much of the upper plumage blotched conspicuously with mahogany color and black; the back (under the overlaying feathers of wings and shoulder region) white, with a blackish patch upon the rump; the tail, showing a blackish field, skirted

No. 54. Adult.

unevenly with white. Under parts white (from the black of the breast backward), and a bar of white crossing wing. Bill nearly black. Legs and toes rich orange red.

Adult female. Similar to adult male, but with the mahogany color and black less positive.

Young. Without mahogany color or black; upper parts unevenly marked instead with brown and yellowish gray; the breast duskily mottled, or showing in a shadowy and imperfect way the markings of the adult, the whole plumage having a very commonplace and washed-out appearance, compared with that of the full-dressed bird; the bill less black; the legs and feet pale orange.

Length eight and a half to nine inches; extent seventeen to eighteen inches. Bill about seven eighths of an inch long.

A "nearly cosmopolitan" species, found chiefly along the sea-coast, but met with also on the shores of large inland waters.

I have eaten this kind several times, but can now only remember my sensations upon one occasion. I tried them a year or two ago on the Jersey coast, and though I was very hungry at the time, they proved altogether too strong for me.

TURNSTONE (so called from its habit of turning over small stones in search of food): **COMMON TURNSTONE** (distinguishing it from *A. melanocephala* of the Pacific coast, a similar but blacker bird, without the mahogany color of this species): **SEA DOTTEREL** (Catesby): **HEBRIDAL SANDPIPER** (Pennant, and Hearne): Hearne tells also of its being known at Hudson's Bay as **WHALE-BIRD**, on account of its "feeding on the carcasses of those animals," and he remarks concerning its flesh: "They are usually very fat, but even when first killed they smell and taste so much like train-oil as to render them by no means pleasing to the palate" (Journey to Northern Ocean, 1769–72, published 1795).

Wilson says (1813): "On the coast of Cape May and Egg Harbor this bird is well known by the name of the **HORSE-FOOT SNIPE**, from its living during the months of May and June almost wholly on the eggs or spawn of the great king crab, called here by the common people the horse-foot." I have made many inquiries along "the coast of Cape May and Egg Harbor," but can nowhere find the latter name so applied (see No. 52), yet in some out-of-the-way corner No. 54 may still be the Horse-foot Snipe as it was in Wilson's time. I will add that this "horse-foot" spawn is very greedily devoured by most of our shore-birds. (See "Horse-foot Marlin" under Nos. 58 and 60.)

Giraud (1844) gives **BEACH-BIRD** as used at Egg Harbor to designate the young; and Mr. William Gaskill, the oldest and perhaps best-informed gunner in the neighborhood of Tuckerton, N. J., tells me that he remembers this as a name formerly used for the species, but not as confined to the young bird. It is not a very distinctive title, and has been given to the Ring-necked Plover—*Ægialitis semipalmata*, Piping Plover—*Æ. meloda*, Sanderling—*Calidris arenaria*, etc.

De Kay, in Zoölogy of New York, 1844, gives **HEART-BIRD** as one of the names applied to the species by "our *gunners*, a class of men who earn a livelihood by shooting birds." This author was evidently afraid that we might confuse the gunner with that helpless but interesting creature "the true sportsman."

In Hallock's Sportsman's Gazetteer, **SAND-RUNNER.**

In Maine at Portland and Pine Point, in Massachusetts at Rowley, Ipswich, Salem, Boston markets, North Scituate, Provincetown, West Barnstable, and Martha's Vineyard, **CHICKEN-PLOVER** and **CHICKEN-BIRD**; at Chatham, **CHICKEN** simply; and referred to in Forest and Stream of September 4, 1873, as **CHICKLING.** Again, in Massachusetts at New Bedford, **RED-LEGS**, and Maynard gives **RED-LEGGED PLOVER** in his Birds of Eastern Massachusetts; at Falmouth, **SPARKED-BACK, STREAKED-BACK**, and **BISHOP PLOVER**; at Nantucket, **CRED-DOCK.** In Connecticut at Saybrook and Lyme, **SEA QUAIL.** On Long Island at Shinnecock Bay, Moriches, Bellport, and Seaford, **BRANT-SNIPE** and **BRANT-BIRD** (see Nos. 53, 60, 61); again at Moriches, **MAGGOT-SNIPE**; at Amityville, **JINNY.** In New Jersey at Manasquan, Pleasantville (Atlantic Co.), Atlantic City, throughout Cape May County, and in Virginia at Eastville, **CALICO-BACK.** Again, in New Jersey at Cape May City, **CALICO-BIRD**; at Pleasantville, above mentioned, **CALICO-JACKET**; at Barnegat, **CHECKERED-SNIPE**; at Tuckerton, **GANNET**;* at Som-

---

* The true Gannets are large fish-devouring sea-fowl of the genus *Sula*.

ers Point, **CHUCKATUCK**; and at St. Augustine, Fla. (to some native gunners at least), **SALT-WATER PARTRIDGE**.
Yarrell tells of its being known in Dorsetshire, England, as
the **VARIEGATED PLOVER**; and the following names appear
in Swainson's Provincial Names of British Birds, 1885: **STANE-
PECKER** (Shetland Isles): **SKIRL CRAKE** (East Lothian, Shetland Isles): **TANGLE PICKER** (Norfolk)—"tangle is a kind of
weed beset with small bladders" (*Gurney*): **STONE RAW** (Ar-
magh): **SEA LARK** (Ireland)—and in the same work under
Sanderling, *C. arenaria*, we find again "Sea Lark (Ireland)."

Our Common Gannet (*S. bassana*), known also as White Gannet, Solan Goose,
etc., measures about three feet in length. Under the head of Royal Tern
(*Sterna maxima*), in Notes on Birds found Breeding on Cobb's Island, Va.
(Bull. Nutt. Ornith. Club, April, 1870), Mr. Bailey says, "called Gannets by the
natives;" and Audubon writes as follows concerning the Wood Ibis (*Tantalus
loculator*): "The Spaniards of East Florida know them by the name of 'Gannets.' . . . At St. Augustine I was induced to take an excursion to visit a
large pond or lake, where I was assured there were Gannets in abundance,
which I might shoot off the trees provided I was careful enough. On asking
the appearance of the Gannets, I was told that they were large white birds,
with wings black at the end, a long neck, and a large sharp bill. The description so far agreeing with that of the Common Gannet or Solan Goose, I
proposed no questions respecting the legs or tail, but went off. Twenty-three
miles, reader, I trudged through the woods, and at last came in view of the
pond; when lo! its borders and the trees around it were covered with Wood
Ibises. Now, as the good people who gave the information spoke according
to their knowledge, and agreeably to their custom of calling the Ibises Gannets, had I not gone to the pond I might have written this day that Gannets
are found in the interior of the woods in the Floridas, that they alight on trees,
etc., which if once published, would in all probability have gone down to future
times through the medium of compilers, and all perhaps without acknowledg-
ment."

No. 55.

Charadrius squatarola.

Sometimes confused with the next species, but differing from it in many ways. This is a *four-toed* plover, having a minute but perfectly distinct hind toe (No. 56 being without this rudimentary appendage). It is also larger, stockier, bigger billed, is a less numerous and more timid species, with louder, fuller note, and is found far more often on the sea-shore proper, upon sandbars, mud flats, and beaches.

No. 55. Breeding Plumage.

Adult in breeding dress. Upper part of head and back of neck white, more or less marked with pale grayish brown; remaining upper parts mottled with white, black, and two shades of brown; tail barred with white and black. Lower part of head and most of the lower plumage plain black (or brownish black), changed to white about vent and root of tail. Bill and legs black.

Adult, and young, in fall and winter. Not now "black-bellied," but a "gray" plover; without the positive contrasts just described; clothed instead with Quaker-like simplicity. Upper parts with neck and portions of breast finely streaked and speckled with grayish brown and white; the upper parts sometimes washed here and there with faint yellow. Remaining under parts white. Bill and legs less black, or grayish in tone.

No. 55.   Fall or Winter Plumage.

Measurements about as follows: length eleven and a quarter to twelve inches; extent twenty-three and a half inches; bill one and a quarter inches.

It should be borne in mind, in this connection and others, that a bird does not change its dress as a snake does its skin, but that while passing from one plumage to another (as in the case of this bird's belly from black to white, and *vice versa*) various combinations are produced.

"Nearly cosmopolitan, but chiefly in the Northern Hemisphere, breeding far north, and migrating south in winter; in America to the West Indies, Brazil, and New Granada" (A. O. U. Check List).

BLACK-BELLIED PLOVER: SWISS PLOVER: WHISTLING PLOVER

(see No. 56): **OX-EYE** (given also to those very common and very small sandpipers, *Tringa minutilla* and *Ereunetes pusillus*, better known as "peeps"): **SWISS SANDPIPER** and **GRAY SANDPIPER** of Pennant, and **GRAY LAPWING** of Swainson and Richardson. Wilson writes: "Called by many gunners along the coast the **BLACK-BELLIED KILLDEER**;" and again: "This bird is known in some parts of the country by the name of the large whistling field plover. It generally makes its first appearance in Pennsylvania late in April; frequents the countries towards the mountains; seems particularly attached to newly ploughed fields, where it forms its nest." Audubon speaks of its breeding "in the mountainous parts of Maryland, Pennsylvania, and Connecticut," and of finding its nests "in the same localities as those of *Totanus bartramius*" (now *Bartramia longicauda*), and he adds that it is known "in Pennsylvania by the name of whistling field plover." Nuttall also calls it "large whistling field plover," and speaks of its being "known to breed from the open grounds of Pennsylvania to the very extremity of the Arctic regions;" and Dr. Lewis, in his American Sportsman, tells of its returning from the South early in May, and soon after retiring to the "high upland districts to breed," and of its being known "at this time more particularly as the old field-plover or whistling plover," and he adds: "A most capital manœuvre, and one adopted by some of our sporting friends in the country, is to approach them in a careless manner, either in an old wagon or cart or on horseback, as they seldom take alarm at a horse or a vehicle of any description." Now No. 55 does not breed in the United States, and Wilson and the rest got it sadly mixed up with the Bartramian Sandpiper, No. 50; and Dr. Lewis's account of the manner in which his birds were pursued is plainly a description of a venerable trick still practised on "field plover" No. 50.

I have found neither this bird (No. 55) nor the following plover (No. 56), sufficiently well known or common enough along the Maine coast from Eastport to Ash Point to have any well-established names. With the exception of a few individuals who shoot over dogs in the brush, the gunners of this section are principally seafowl-shooters who know very little, and care less, about shore-birds.*

At Ash Point (near Rockland), Me., Seaford (Hempstead), L. I., and Barnegat, N. J., **GRAY PLOVER** (see No. 50). In Maine at Bath, Portland, and Pine Point, at Portsmouth, N. H., in Massachusetts at Ipswich, Salem, North Scituate, Provincetown, West Barnstable, Chatham, and New Bedford, at Stratford, Conn., and Shinnecock Bay, L. I., **BEETLE-HEAD**; at Eastville, Va., **BEETLE** simply.   Again, at Bath and Portland, **CHUCKLE-HEAD.**   At North Plymouth, Mass., **BOTTLE-HEAD.**   On Long Island at Shinnecock Bay, and in New Jersey at Manasquan, Tuckerton, Atlantic City, Somers Point, Cape May C. H., and Cape May City, **BULL-HEAD** (see No. 56); at Stonington, Conn., **BULL-HEAD PLOVER.**   "In the Eastern States," Audubon wrote, "as well as in Kentucky, it is called the Bull-head; but in the South its most common appellation is Black-bellied Plover." In New Jersey at Pleasantville (Atlantic Co.), and Atlantic City, **HOLLOW-HEAD**; and again at Pleasantville, **OWL-HEAD.**   At Pine Point, Me., Portsmouth, N. H., in Massachusetts at Provincetown, West Barnstable, Chatham, New Bedford, and Falmouth, and at Stratford, Conn., **BLACK-BREAST** (see Nos. 53, 56).   On Long Island at Moriches, Bellport, and Seaford, and

---

* The term "shore-birds," as commonly used, means such species as the curlews, plovers, sandpipers, etc., these being also termed "bay-birds" by many; and Wilson wrote, while describing the red-backed sandpiper, No. 53: "This is one of the most numerous of our *strand birds* as they are usually called." Shore gunners very naturally associate birds of this kind almost wholly with the beaches and meadows that border the sea, yet the same species are to be met with, as well, away back upon the prairies of the interior, particularly during the vernal migrations. Other titles used to designate these birds collectively I have noted as follows: At Ash Point, Me., "sand-birds." In Maine at Bath and Kennebunk, at Portsmouth, N. H., in Massachusetts at Rowley, Salem, North Scituate, Provincetown, North Plymouth, and Barnstable, and at Newport, R. I., "marsh-birds." On Long Island, and in its vicinity, "bay-snipe" and "shore-snipe." These are the only notes of the kind that I can find among my memoranda, and I will not attempt to continue the list from memory.

in New Jersey at Barnegat, Tuckerton, and Cape May City,
BLACK-BREAST PLOVER.

In numerous localities gunners divide this species; for example, the name Beetle-head is commonly restricted to birds in
autumn dress, these being regarded as distinct from the Black-
breasts.

Known also at West Barnstable, Mass., as **MAY COCK**; and in
that interesting pamphlet about Shore Birds—No. 1, of the Forest
and Stream series, 1881, we find the following: "On the coast
of Virginia about Cobb's Island, the name of **PILOT** has been
given, as it is always seen leading the large flights of birds
which the rising tides drive from the shoals and oyster rocks,
and it is supposed to direct the flocks 'to pastures new.' This,
however, is not the case. It is the fastest flying bird of all the
bay snipe, and it cannot fly slow enough for the other species."·

Mr. Swainson writes, in his Provincial Names of British Birds,
1885: "Its habit of frequenting the sea-shore has obtained for
it the names **SEA PLOVER**: **SEA COCK** (Waterford): **STRAND
PLOVER** (Cork): **MUD PLOVER, STONE PLOVER** (North and
South Ireland): **ROCK PLOVER** (Wexford)."

No. 56.

Charadrius dominicus.

Sometimes confused with No. 55, but smaller, with a more slender bill and three-toed (the latter species having a minute hind toe); much more of an upland bird; partial to rather barren or closely cropped fields, particularly to patches that have been recently burned over.

Adult in breeding dress. Forehead and stripe over eye white; upper parts generally brownish black speckled with yellow and white, the tail grayish brown with lighter markings. Under parts, including the lower part of head, rich brownish black. Bill nearly black. Legs dark bluish gray.

Adult at beginning of autumn (as we kill it on its journey southward). Under parts white or ashy white, blotched irregu-

No. 56. "Muddy-breast" Plumage.

13

larly with brown or brownish black* ("Muddy-breast" now); top of head dark ; the stripe over the eye light, but not white; remaining upper plumage much as in the breeding dress, but more dull in tone.

Young (reaching us, the greater part of them, a little later than the adults). Upper parts dark brown and brownish gray speckled with yellow and dull white; the yellow showing noticeably on the crown, nape, and lower back (all the speckles of the lower back yellow); head and neck streakily marked, the

No. 56. Young.

neck lighter in color than top of head or back ; sides of head and all around base of bill white or whitish streakily marked with dusky. Under parts of plumage and sides of neck dull white freckled with brownish gray, the chin pure white, the brownish gray markings much paler on the lower surface of body, particularly about the abdomen, and nearly or wholly disappearing in neighborhood of vent.

---

* These blackish markings wholly disappearing in winter, at which time the plumage throughout is practically that of the young bird.

This species closely resembles the Golden Plover of Europe (*C. apricarius*), but in our bird the lining of the wings is gray, while in the European it is white.

Measurements about as follows: length ten and a half inches; extent twenty-two inches; bill one inch long.

A delicious bird for the table, and everywhere regarded as such; breeding in the Arctic regions, and migrating in large flocks to the southern extremity of South America.

AMERICAN GOLDEN PLOVER: COMMON PLOVER: WHISTLING PLOVER (see No. 55): GOLDEN-BACK: BULL-HEAD (see No. 55). Edwards, 1750, describing it under the name of the SPOTTED PLOVER, says: "This bird was brought from Hudson's Bay by Mr. Isham. I suppose when it is living it has a bright shining eye, because I find by my friend Mr. Isham's account that the English settled in Hudson's Bay call it the HAWK'S EYE." Wilson, citing Pennant, credits the Black-bellied Plover, No. 55, with this name Hawk's-eye, believing Pennant's "Alwargrim Plover" (Arctic Zoölogy, p. 483, No. 398) to be the latter species.

In Maine at Ash Point, FIELD-BIRD; at Bath, THREE-TOED PLOVER; and a venerable gunner of Bath, Mr. Samuel Foote, remembers this latter name as so applied in his early childhood. At Portland, Me., and in Massachusetts at Rowley, Ipswich, North Scituate, North Truro, and North Plymouth, and at Stonington, Conn., BLACK-BREAST (see Nos. 53, 55). At Portland, Me., Bellport, L. I., and Stratford, Conn., GOLDEN PLOVER. At Portsmouth, N. H., and in Massachusetts at Salem and Chatham, GREEN PLOVER. At Provincetown, Mass., and Moriches, L. I., GREEN-BACK. In Massachusetts at Rowley, BRASS-BACK; at Ipswich and North Scituate, PALE-BREAST; at Provincetown, New Bedford, and Chatham, PALE-BELLY; these last two names being applied only to the young birds, which are regarded by many as a distinct species or variety. At West Barnstable and New Bedford (Mass.), and Newport, R. I., GREEN-HEAD; and to the old people of West Barnstable, PASTURE-BIRD (a name now seldom heard, but used there by every one until fifteen or

twenty years ago). At Newport, R. I., **MUDDY-BREAST**; and at Seaford, L. I., **FROST-BIRD.***
Mr. Henry P. Ives, of Salem, tells of its being known as **TROUT-BIRD** at Hamilton, Mass. Mr. Browne records **SQUEALER** in his list of gunners' names at Plymouth Bay (Forest and Stream, November 9, 1876). Mr. John Murdoch (Forest and Stream, December 9, 1886) speaks of hearing it called **TOAD-HEAD** on Cape Cod; stating that most of his "shore-bird nomenclature for Cape Cod was learned in the town of Orleans in the seasons of 1869–72, and chiefly from the older generation of gunners." Mr. M. A. Howell, Jr., writes (Forest and Stream, March 1, 1877): "From the regularity of the visits of these birds in former years to the sand bars of the upper Illinois and Kankakee, they have been called by the resident shooters, **KANKAKEE BAR PLOVER.**" Mr. Warren Hapgood, in Forest and Stream Shore Bird pamphlet, 1881, speaks of its being known in the West as **PRAIRIE PIGEON** (see No. 50); and writes in reply to inquiries of mine that he has forgotten just where he heard the name in use; but he adds, "It was common talk when I was in Iowa, before the article was written, that the earlier settlers were annoyed by these birds, which, in the absence of a better name, they called Prairie Pigeons."

---

* Herbert, in his Field Sports, credits the Bartramian Sandpiper, No. 50, with this name Frost-bird, but later on applies it correctly to No. 56.

No. 57.

Numenius longirostris.

Plumage brownish buff or cinnamon brown, nearly plain below, the upper parts mottled and barred with dark brown, the markings more streaky upon the head and neck. Bill blackish, changing to flesh color below about base. Legs bluish gray.

Measurements as follows: length about twenty-five inches; extent thirty-eight to forty inches; bill varying in length from five to eight inches.

No. 57.

Range : " Temperate North America, migrating south to Guatemala and the West Indies. Breeds in the South Atlantic States, and in the interior through most of its North American range" (A. O. U. Check List).

Not now a common species in New England or north of New Jersey, and noticeably less common along the shores of New Jersey, Delaware, Maryland, and Virginia, than in former years. In South Carolina and southward, and in interior parts of the country, it is met with in goodly numbers.

SICKLE-BILL CURLEW, or SICKLE-BILL; very generally known as such, or as the BIG CURLEW, along the coast as far south as Maryland at least, though otherwise designated as shown in the following list.

In Massachusetts at Rowley and New Bedford, HEN CUR-LEW, or OLD-HEN CURLEW. At Shinnecock Bay, Moriches, Bellport, and many other Long Island localities, this is the only CURLEW ; the Hudsonian, No. 58, being the " Jack," and the Es-kimo, No. 59, the " Fute " or " Doe-bird " (see " Curlew " as ap-plied to the Marbled Godwit, No. 60). In New Jersey at Tucker-ton, LONG-BILLED CURLEW; at Absecum, Pleasantville (Atlan-tic Co.), and Somers Point, BUZZARD CURLEW (its flight re-sembling that of a turkey-buzzard) ; known also at Pleasantville to some of the gunners as SMOKER or OLD SMOKER (the bill curving downward like the stem of a pipe, and the enlarge-ment at the end answering for the bowl) ; again at Pleasant-ville, LOUSY-BILL (the bird being frequently found infested with lice) ; at Cape May C. H., MOWYER (an old-fashioned word mean-ing one who mows). To many gunners along the shores of South Carolina and Georgia, and at St. Augustine, Fla., it is the SPAN-ISH CURLEW — this name being given in books to the White Ibis, *Guara alba.* Mr. Ridgway (in Survey of Fortieth Parallel, 1877) speaks of its being " called SNIPE by the people of the Salt Lake Valley ;" and also of its being " particularly abundant along the southern shore of the Great Salt Lake, and some of the larger islands."

In Hallock's Gazetteer (1877), SABRE-BILL.

Numenius hudsonicus.

Upper parts brown, the feathers edged and otherwise marked with whitish; general appearance similar to species No. 57, but paler in tone and more gray. Throat and belly whitish with some pale buff. Crown of the head blackish brown, divided in the middle by a white, or nearly white, streak running from the

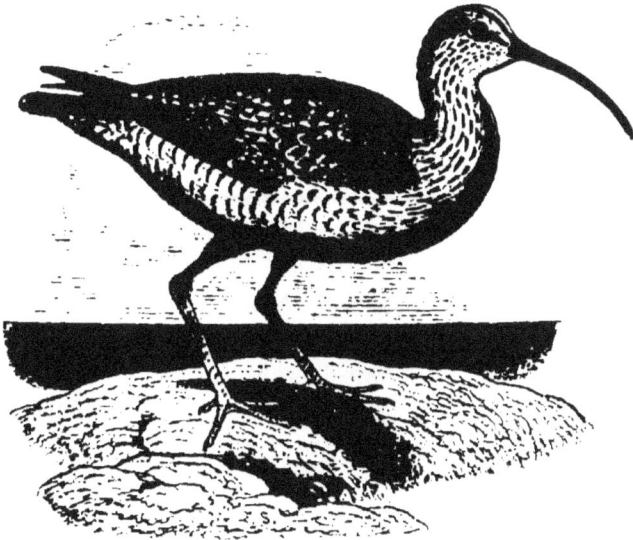

No. 58.

bill backward; also a dark stripe running from the bill along the side of the head—the head alone easily distinguishing this bird from our other curlews, Nos. 57 and 59. Bill black or blackish, flesh-colored below near base. Legs blue or bluish.

Length seventeen to eighteen inches; extent about thirty-two inches; bill varying in length, say from two and three quarter to three and three quarter inches.

Range, according to A. O. U. Check List: "All of North and South America, including the West Indies; breeds in the high North, and winters chiefly south of the United States."

HUDSONIAN CURLEW: called by Wilson and Nuttall, ESQUI-MAUX CURLEW (see Esquimaux Curlew proper, No. 59), these authors following the lead of Pennant, who, according to Fauna Boreali - Americana, had "misapplied Mr. Hutchins's notes." Pennant also refers to the present species as the ESQUIMAUX WHIMBREL (because of its resemblance to the European curlew, *N. phæopus*, which is known as Whimbrel).

At Pine Point, Me. (I have no notes of hearing the gunners name it north of this place), and in Massachusetts at Province-town and Chatham, JACK CURLEW.· On Long Island at Shinne-cock Bay, Bellport, and Seaford, JACK, and Mr. William Dutcher, in Forest and Stream, August 5, 1886, speaks of its being called "almost universally on Long Island, Jack." Not *Jack - curlew* be it understood, the only surname ever added in that locality being "snipe;" all the waders are "snipe" or "bay-snipe" there. In New Jersey at Barnegat, SMALL CURLEW;* at Tuckerton and Cape May City, SHORT-BILLED CURLEW; at Pleasantville (Atlantic Co.), and Cape May C. H., MARLIN (see No. 60); again, at Pleasantville and at Somers Point, CROOKED-BILLED MARLIN; in last-named locality, HOOK-BILLED MARLIN; and at Atlantic City, HORSE-FOOT MARLIN, because of its fondness for the spawn of that big crustacean known as "horse-foot," "horseshoe," "king-crab," etc., but, as elsewhere remarked, this food is re-garded as very desirable by most of our shore birds. While the Atlantic City gunners claim that No. 58 is the species to which the latter name has been always applied in their region, the

---

* This is the most common curlew along the coast of New Jersey, and the most common (I speak from my own experience) along the coasts of Dela-ware, Maryland, Virginia, and North Carolina, and gunners in that part of the country know very little about the Eskimo, No. 59, to which species the name Small Curlew more appropriately belongs.

" Horse-foot Marlin " of the Somers Point gunners (same county) is the Marbled Godwit, No. 60.

At Eastville, Va., **STRIPED-HEAD**, in which vicinity the species is exceedingly numerous during its vernal migration; arriving from the South at the beginning of May, and congregating in enormous flocks in and about the broad marshes. I originally intended to print the number locally reported as killed on New.Marsh, between Cobb's Island and the mainland, by one discharge of a gun held by Nathan Cobb (familiarly known as " Big Nathan "), but my best friends strongly advise me not to do so. While on my way through these marshes in the spring of 1885—frightening into the air clouds of these big birds, more in a minute than I had seen before in my whole life— it impressed me oddly to hear my old boatman complaining over a yearly decrease. I forced him to confess, however, at one point where the birds were particularly crowded, that " a right-smart of curlews " was still left.

For the name "Doe-bird" with which this species has been credited, see note (†) under No. 59.

No. 59.

Numenius borealis.

This bird may be briefly described by comparing it with the other curlews. General markings and coloration very similar to No. 57; prevailing tone of plumage warmer or more reddish than that of No. 58, and bill much slenderer, as well as shorter;

No. 59.

differing also from No. 58 in having no white stripe on top of the head. Bill blackish, flesh-colored beneath about base. Legs grayish blue.

Measurements about as follows: length fourteen inches; extent twenty-eight inches; bill two and a quarter inches.

Range, as given in A. O. U. Check List: "Eastern Province
of North America, breeding in the Arctic regions, and migrating
south to the southern extremity of South America."
A better bird for the table than either of the other curlews;
much more of an upland species; very fond of berries and grass-
hoppers, and frequently found in the company of Golden Plover
(migrating from the North at about the same time). Its visits
to us are more irregular and less protracted than those of No.
58, with which it has been sometimes confounded.

ESQUIMAUX CURLEW (now written Eskimo Curlew. See
No. 58): **LITTLE CURLEW**, and **SMALL CURLEW** (again see No.
58); and Nuttall (1834), having applied the name Esquimaux
Curlew to No. 58, called this the **SMALL ESQUIMAUX CURLEW.**
At Pine Point, Me., in Massachusetts at Ipswich, Salem,
North Scituate, Provincetown, North Truro, North Plymouth,
West Barnstable, Chatham,* and Nantucket, and at Moriches,
L. I., **DOE-BIRD**† (written also Dough-bird). At Stratford, Conn.,

---

* This species appears on the more eastern uplands of Cape Cod the last
of August or during the early days of September, and if severe easterly
storms prevail, it arrives in very large numbers. The Hudsonian, No. 58, is
far less numerous here, and the Long-billed, No. 57, may now be called rare.
No. 59 is a great favorite with Boston epicures, and the gunners get from
seventy-five cents to a dollar apiece for them; as a table dainty I consider
them superior to all other birds, but they should hang with the feathers on, in
a shady, breezy place, for four or five days before being cooked.

† Other species have been credited with this name, but I do not remember ever hearing it in actual use for any bird but the Eskimo, to which it
now, at least, most certainly belongs. Nuttall, who was, of course, thinking
more of the birds themselves than of their common names, and who did not
perhaps fully realize the importance of such names as bearing upon the science
itself, tells us that the three species Nos. 58, 59, and 60 were included "under the general name of Doe-birds." It is hard to believe that the gunners
ever mixed up these birds so indiscriminately. It is possible, of course, that
some used the name as we use "bay-bird," "sea-coot," etc., but I am inclined
to think that "Doe-bird" was used then by intelligent gunners, as it is now,
for No. 59 only. Later writers—more or less influenced perhaps by Nuttall's
testimony—must also be referred to in this connection. De Kay, in Zoölogy
of New York, credits Nos. 59 and 60 with this name. Samuels, in Ornithology

and on Long Island at Shinnecock Bay and Seaford (Hemp-stead), **FUTE.**

In Water Birds of North America we read of its being known to " Southern sportsmen " as Jack Curlew and Short-billed Cur-lew, and that " it is said to reach the Middle States from the South early in the spring, remaining only a short time, feeding in the salt-marshes and on the mud-flats;" and again, that the Hudsonian Curlew is " generally known to sportsmen " by these two names. Both these items were designed, perhaps, for a place under the head of the last-named species (No. 58), to which the names belong and the description applies.

---

of New England gives it as an alias of No. 60, mentions it under the head of No. 59 only in a quotation which he makes from Nuttall, and speaks of No. 61 as " called by the gunners the Smaller Doe-bird."

No. 60.

Limosa fedoa.

Prevailing tone pale reddish cinnamon ; closely variegated above with dusky brown; dusky markings sometimes about the breast and sides; the lighter tints of the plumage having an

No. 60.

occasional pinkish cast. Bill (curved slightly upward as in picture) flesh colored from the base more than half-way to tip, the remainder blackish brown. Legs dark slate color.

Size varying according to different authors, about as follows:

length from sixteen to twenty-two inches; extent thirty to forty inches; bill three and a half to five and a half inches; and Wilson describes the bill as "nearly six inches in length." I have measured only one freshly killed specimen; its measurements were: length twenty-one and a quarter inches; extent thirty-one inches; bill three and three quarter inches.

Range, as given in A. O. U. Check List: "North America, breeding in the interior (Missouri region and northward), migrating in winter southward to Central America and Cuba."

MARBLED GODWIT: GREAT MARBLED GODWIT: GREAT GODWIT: AMERICAN GODWIT: GREATER AMERICAN GODWIT.* Wilson (1813) speaks of its being sometimes called **RED CURLEW** by "our gunners;" and Maynard, in Birds of Eastern Massachusetts, 1870, records **BADGER-BIRD** and **BRANT-BIRD** (see Nos. 53, 54, 61).

I have but one note of hearing this species named between New Brunswick and Rhode Island, where it is too little known to bear any very well-established names. An old gunner at Salem, Mass., to whom I showed a stuffed specimen, said, "We call that a curlew here."

At Newport, R. I., **COMMON MARLIN**; at Shinnecock Bay, L. I., **RED MARLIN** (and referred to by this latter name in an article about shooting near Barnegat Light, N. J., the communication being headed Snipe at Forked River—Forest and Stream, October 3, 1878). On Long Island at Moriches, Bellport, and Seaford, in New Jersey at Manasquan, Barnegat, Tuckerton,

---

* The name godwit is probably from the Anglo-Saxon *god*, good, and *wiht* or *wihta*, creature, animal, wight. A good bird to eat, in other words. We read in Dr. Thomas Moufet's Health's Improvement, "corrected and enlarged" by Chr. Bennet, 1655: "A fat godwit is so fine and light meat, that noblemen, yea, and merchants too, by your leave, stick not to buy them at four nobles a dozen." In Hearne's Journey to the Northern Ocean, 1795, the name is printed "godwait;" and Dr. Merriam refers as follows (1877) to the spelling in Rev. J. H. Linsley's Catalogue of Connecticut Birds, 1843: "The good old preacher in speaking of these birds could not take his Lord's name in vain on so slight a provocation, hence he called them '*good*wits.'"

and Cape May City, and at Eastville, Va., **MARLIN** (see No. 58);
and Mr. William Dutcher mentions it as "**BROWN MARLIN** of
the Long Island gunners," Auk, October, 1886.   The name Marlin comes from a resemblance in the bird's bill to the old-fashioned marline-spike, which was more or less curved in shape.

In New Jersey at Pleasantville (Atlantic Co.), Townsend In-
let, Cape May C. H., and Cape May City, **SPIKE-BILL**, and less
frequently, **SPIKE-BILLED CURLEW**.   At Atlantic City, N. J.,
Eastville, Va., to some at Morehead,* N. C., and in the vicinity
of Charleston, S. C., **STRAIGHT-BILLED CURLEW**; but more commonly termed in the last two localities, **CURLEW** simply (the true
curlews, genus *Numenius,* being generally referred to as the
"crooked-billed curlews").   Latham (1785) tells of its being
known as "curlew" at Hudson's Bay, and I have before spoken
of hearing it so termed at Salem, Mass.

At Somers Point, N. J., **HORSE-FOOT MARLIN** (see No. 58).

For the name Doe-bird, with which the species has been credited, see note (†) under No. 59.

---

* I killed near Morehead, December 20th, one of the specimens from which
my description was taken. The species is quite common there, though not often
seen so late in the year.

No. 61.

Limosa hæmastica.

Summer plumage. Back and wings grayish and blackish brown with rather angular pale tan markings; tail-feathers black (or brownish black) with ends narrowly tipped with white; lower back nearly plain dark brown separated from the

No. 61.   Summer Plumage.

black of the tail by a broad white patch ("Spot-rump"); head and neck streakily marked with pale gray and blackish; the longer wing-feathers (primaries) deep brown with white shafts and touches of white about their bases. Under parts reddish brown or chestnut barred with narrow dusky and whitish lines, the markings much broader and more conspicuous behind or in the neighborhood of tail; the reddish or chestnut tint continuing, up faintly to a whitish throat; the lining of the wings chiefly sooty brown. Bill flesh color, becoming brownish black at the end. Legs and feet slate color.

Winter plumage. Above light warm gray, nearly plain, with wings and tail about as in summer; the white rump still conspicuous. Under parts warm grayish white without noticeable markings, and becoming purer white behind. Bill and legs as in summer.

Immature birds and adults passing from one plumage to another, of course show intermediate tints and markings.

Length fourteen and a half to seventeen inches; extent twenty-six to twenty-nine inches; bill (curving slightly upward as in picture) two and three quarters to three and a half inches.

Range, as given in A. O. U. Check List, 1886: Eastern North America, and the whole of Middle and South America. Breeds only in the high North.

HUDSONIAN GODWIT: AMERICAN BLACK-TAILED GODWIT: RED-BREASTED GODWIT: ROSE-BREASTED GODWIT: BAY-BREASTED GODWIT.

I have failed to fall in with this bird on the coast of Maine, and none of the many gunners conversed with there are at all familiar with it.

In Mr. Everett Smith's Birds of Maine,* **BRANT-BIRD** (see Nos. 53, 54, 60). In Massachusetts at Rowley, Salem, Boston markets, Provincetown, West Barnstable, and New Bedford, **GOOSE-BIRD**; at Ipswich and Salem, **BLACK-TAIL**; at North Scituate, Provincetown, and Chatham, **SPOT-RUMP**; at West Barnstable, **WHITE-RUMP**. I know of no other part of the United States where this species can be more surely found during its migrations than upon certain portions of the Massachu-

---

* Published in *Forest and Stream*, 1882–83.

14

setts coast. Though in no part of the country is it a common species, so far as I can ascertain.

At Newport, R. I., at all places visited on Long Island, in New Jersey at Manasquan, Barnegat, Atlantic City, and Somers Point, and on Cobb's Island, Va., **RING-TAILED MARLIN.** Again, in New Jersey at Pleasantville (Atlantic Co.), **CAROLINA WILLET;** at Somers Point, **FIELD MARLIN.**

(For the name "Smaller Doe-bird," with which this species has been credited, see note (†) under No. 59.)

INDEX.

THE END.

www.ingramcontent.com/pod-product-compliance
Lightning Source LLC
Chambersburg PA
CBHW021701210326
41599CB00013B/1480